D1689835

An Engineer Imagines

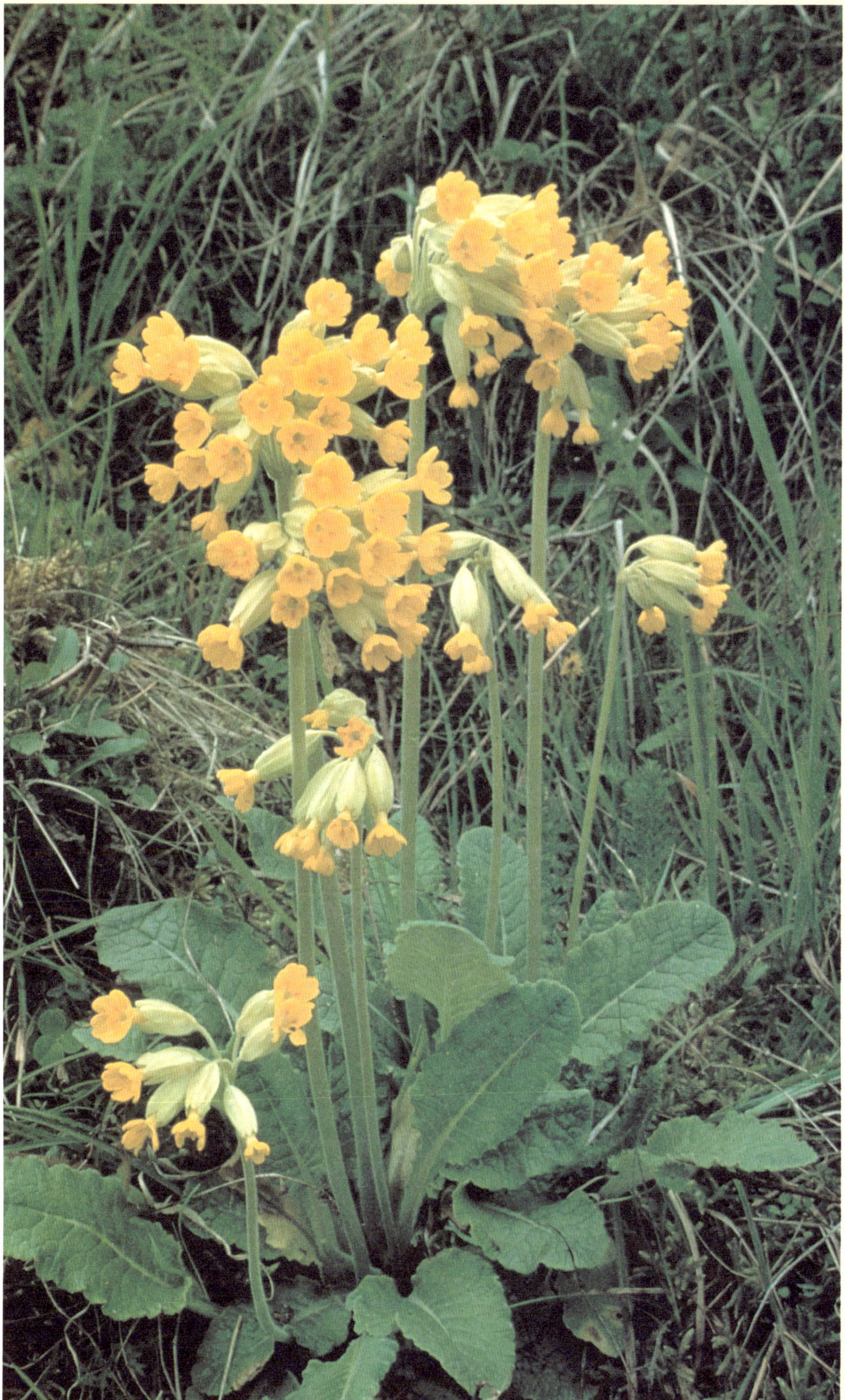

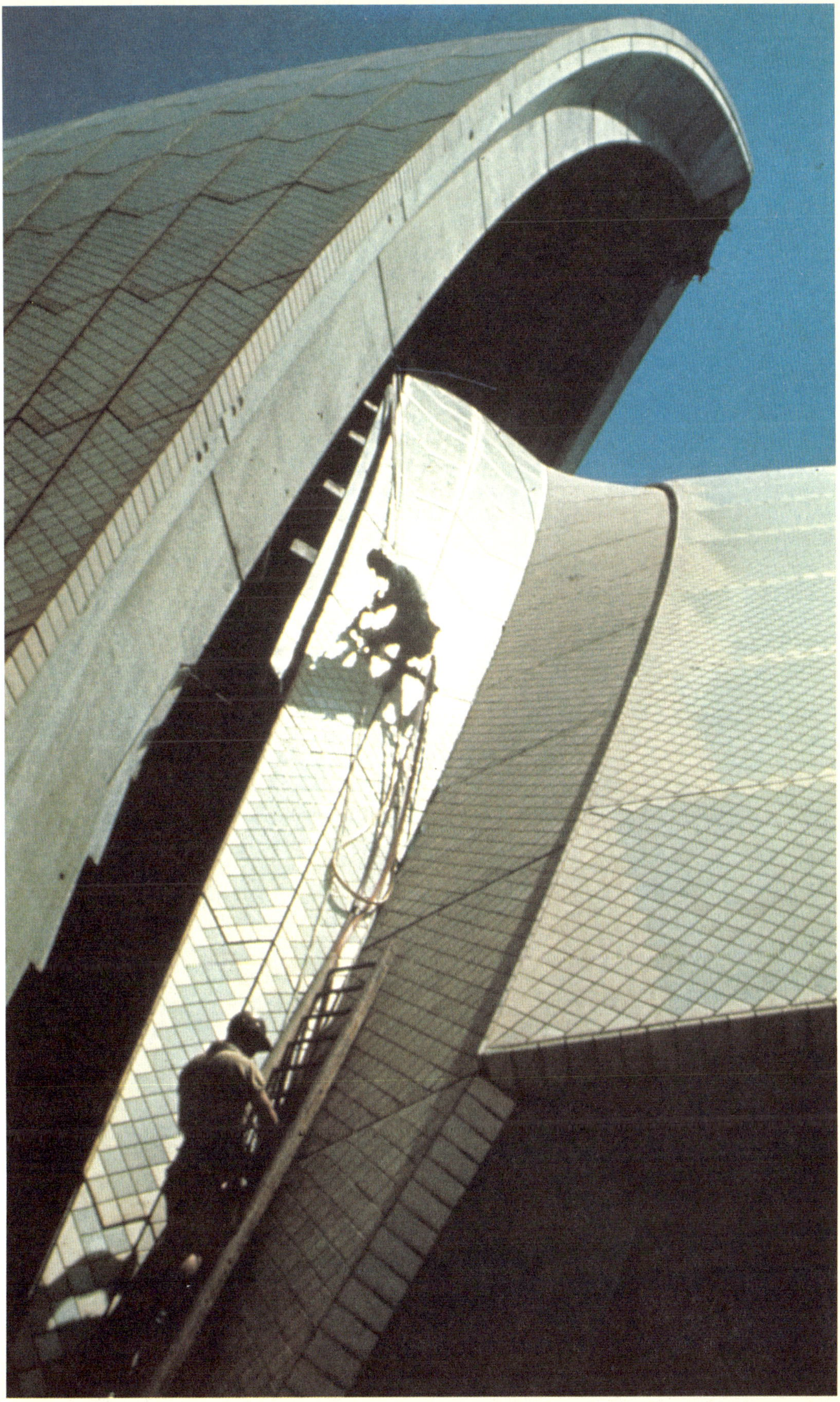

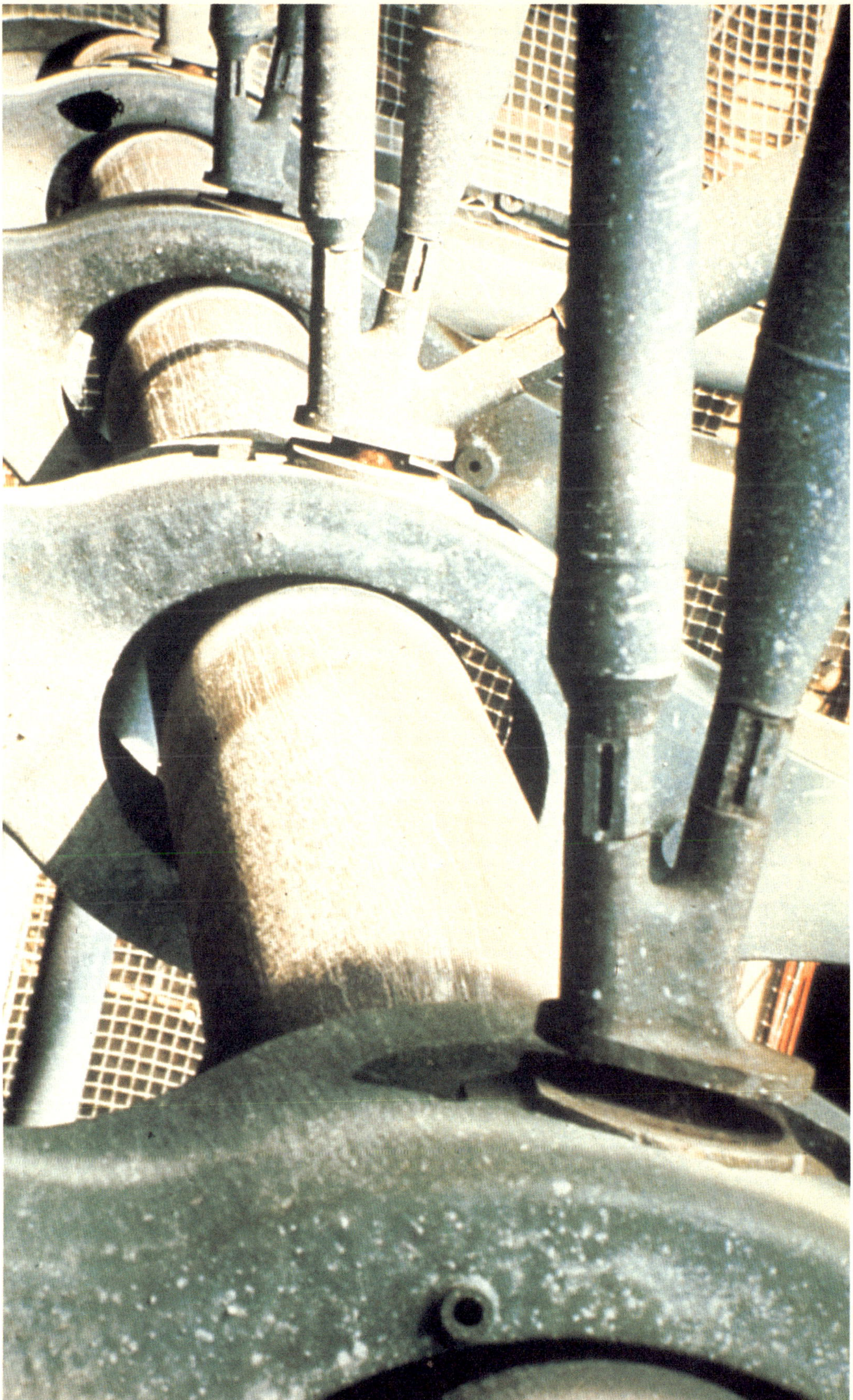

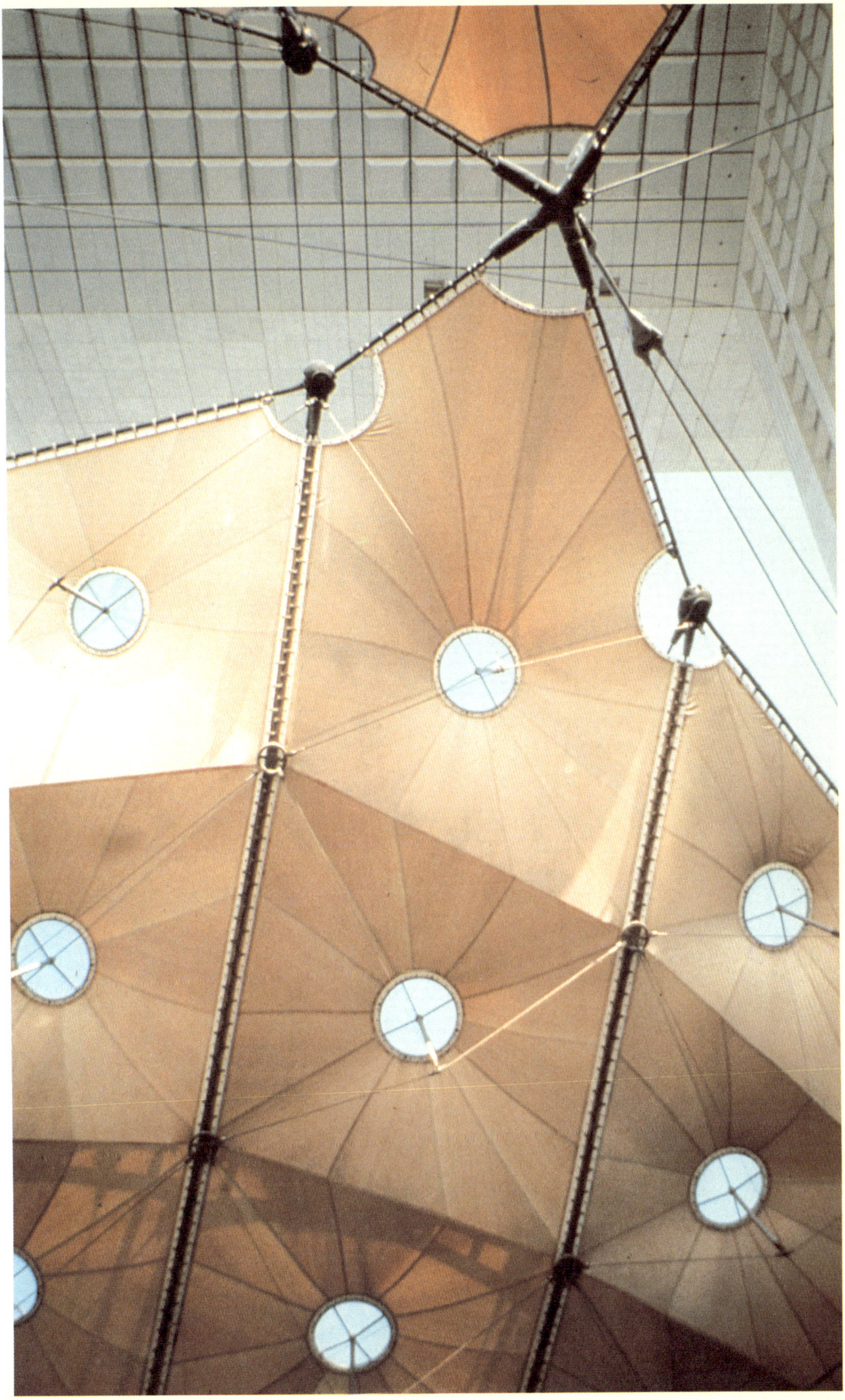

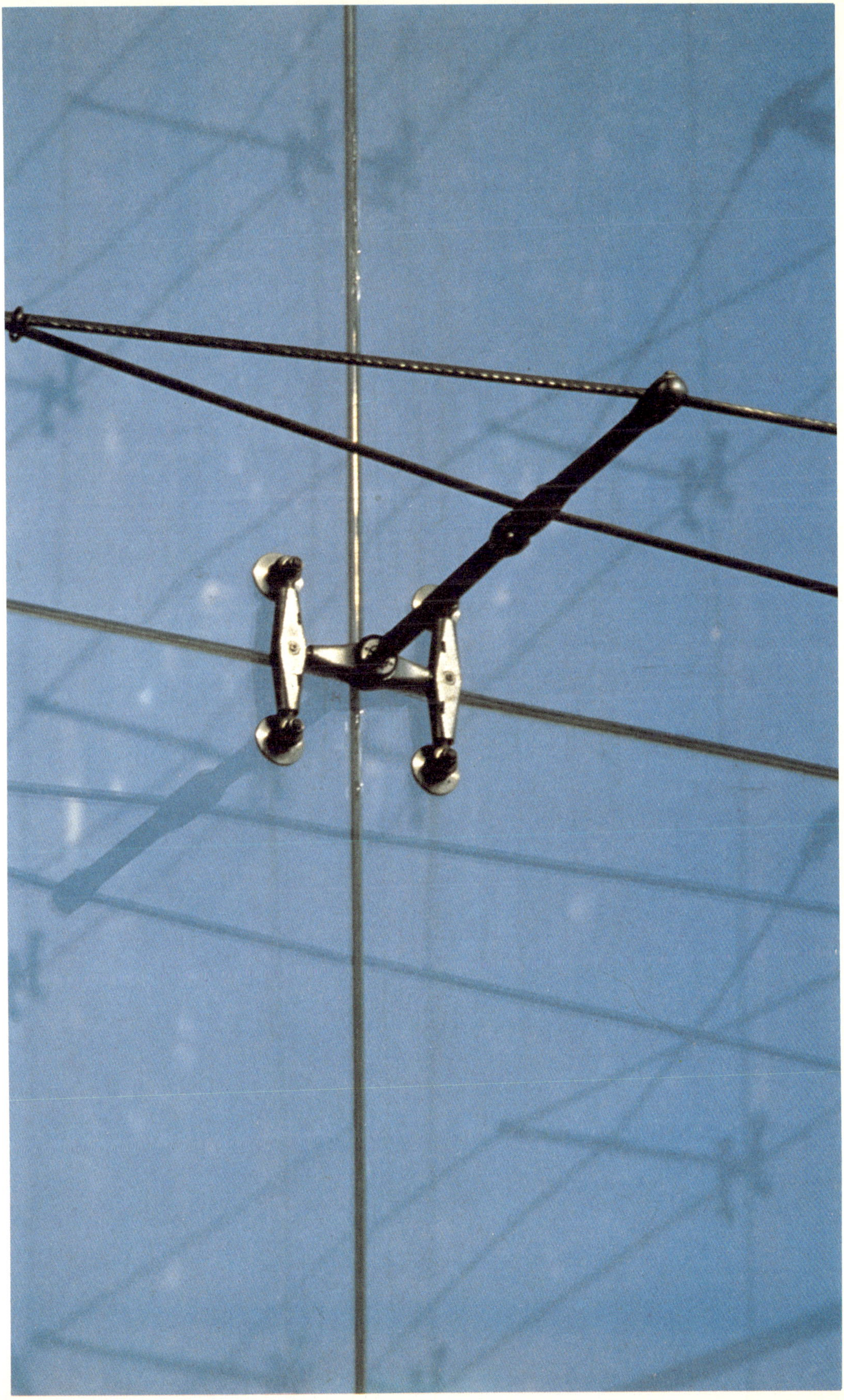

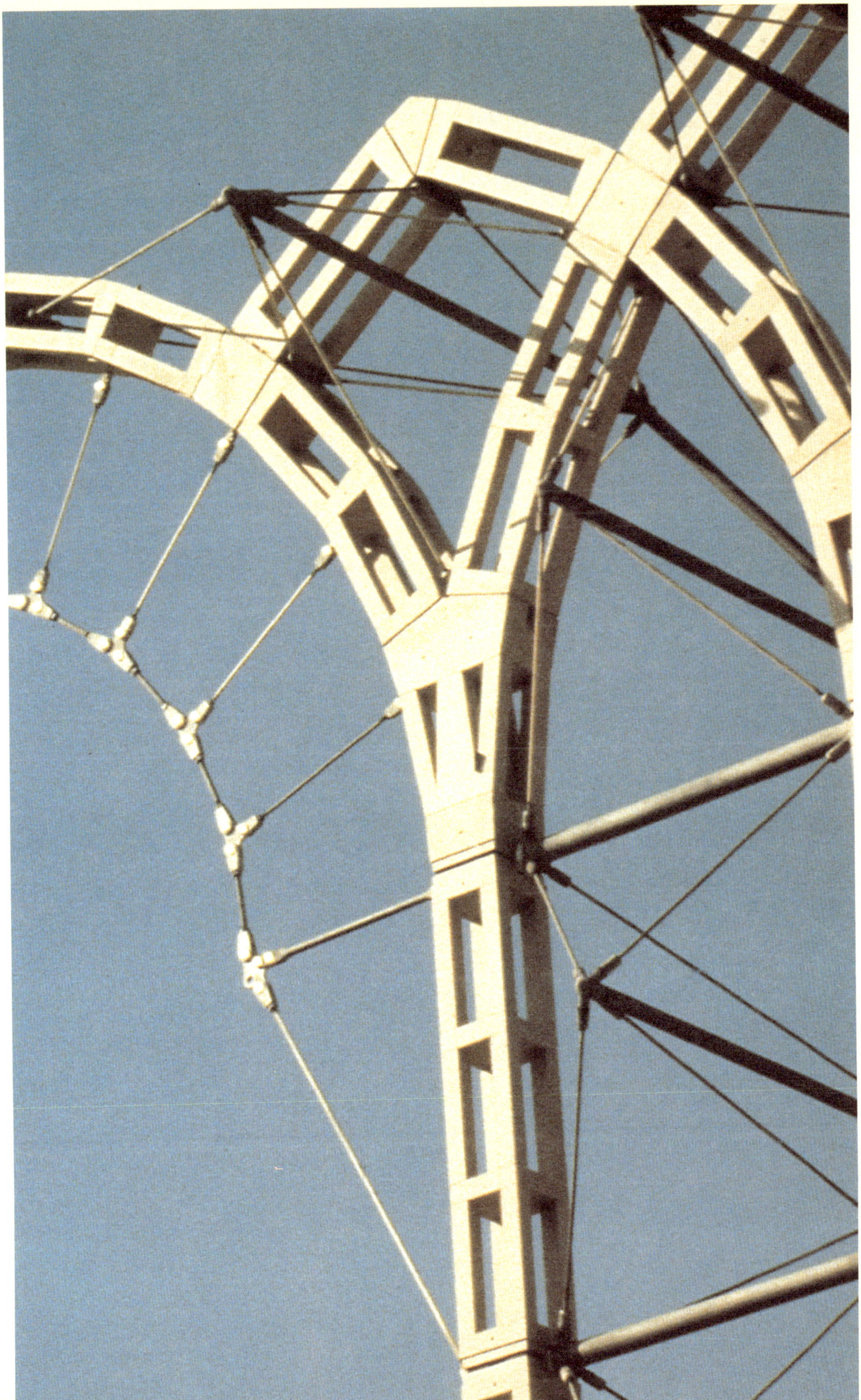

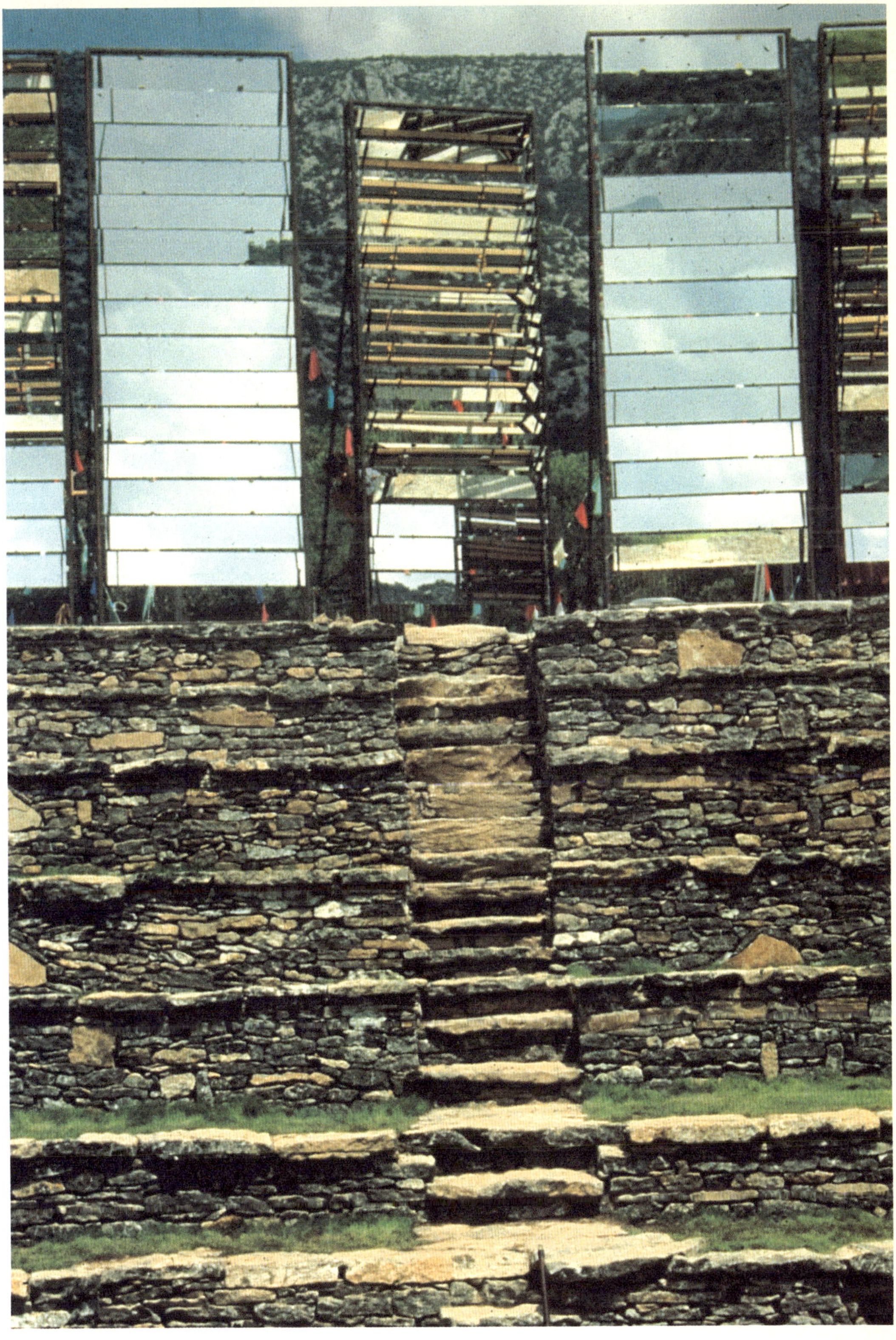

An Engineer Imagines

Peter Rice

Artemis
London Zurich Munich

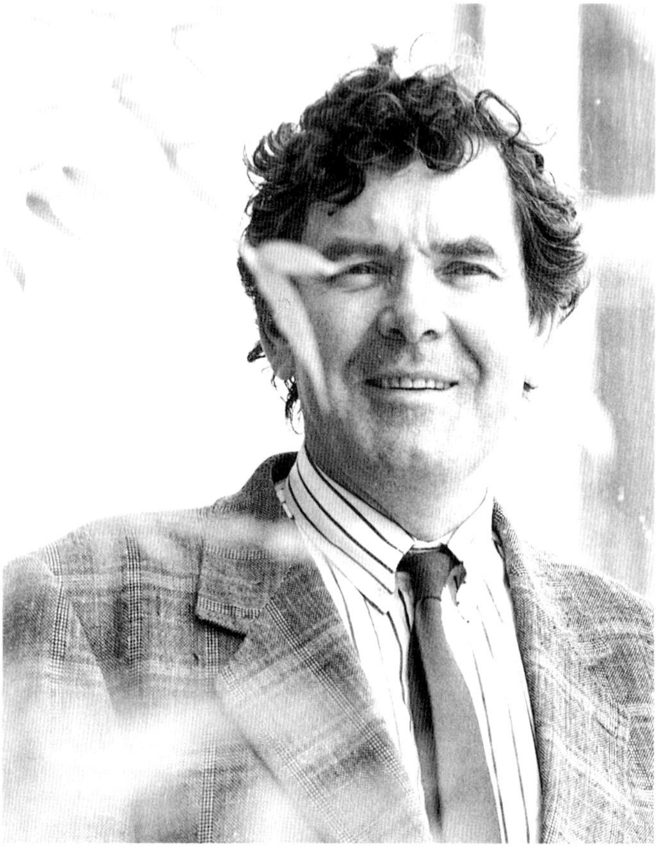

Preface

Many times during Peter's career he was approached to write a book about his work. Unfortunately it wasn't until he became ill that he found the time to do so. He always had a very specific idea about what sort of book he wanted to create – not a textbook about how to solve technical problems but a personal account of the joy and enthusiasm he took from his profession. More than anything Peter wanted to communicate the limitless potential of the engineer's role in pushing back the frontiers of the built environment. Consequently it was very important to him that *An Engineer Imagines* should be accessible to all and not just to an architectural or engineering élite.

To achieve this we know Peter was greatly helped by Barbara-Ann Campbell whose editorial criticism and assistance during the writing of the book proved invaluable. During preparation of the book for publication she has been assisted by Gilly Graham, John McMinn, and Hugh Dutton, with editorial support, discussion with publishers, design and preparation of the manuscript and illustrations. We would also like to thank Caroline Wayne, Peter's personal secretary, for her secretarial support and assistance over many years.

As well as to those mentioned above, the family would also like to say a heartfelt thank you to all those people whose support, encouragement and friendship made it possible for Peter to complete *An Engineer Imagines* despite the emotional and physical pressures of his illness. They are:
Rudolph Archer, Katie Brown, Joy Clarke, Kitty Gibney and family, Lex Hall, Robin Jarossi, Paul Jenkins, Frank Moore, Peggy Muncer, Maurice Rice and family, Paul Rudin, Deborah Sutherland, Chrissie Watson; Jonathan Watson and family; Paul Andreu; Jack Zunz, Tom Barker, Bob Emmerson, Stuart Irons, John Martin, Duncan Michael, John Thornton, David Glover, Sophie Le Bourva, Alistair Lenczner, Andy Sedgwick, Jane Wernick and all those at the Ove Arup Partnership; Humbert and Viviane Camerlo; Michael Dowd; Paul Winkler and all those at the Menil Foundation; Renzo and Milly Piano, Shunji Ishida and all those at the Renzo Piano Building Workshop; Martin Francis, Henry Bardsley, Huguette Boutin and all those at RFR; Ian Ritchie and Jocelyn van den Bossche; Richard and Ruthie Rogers; Frank Stella; Monseigneur Vilnet; John Young; Michael Rose; Dr Michael Weir and all those at the Bristol Cancer Help Centre; Dr Dasgupta; Jan Souridis; Dreas

Reyneke; Mr Henry Marsh, Dr Brada and all those at the Royal Marsden Hospital, London.

Although Peter was unable to write these acknowledgments himself, he wrote a poem in May 1992 which we think expresses how much the support of his friends meant to him, especially in the last year of his life. We would like to dedicate 'Time to Go' to all the close-knit friends mentioned above and those whom we may have inadvertently missed out.

> Tonight I know that I will leave
> The fight is done
> And slowly as I say goodbye
> I need to speak
> And say my love for close-knit friends.
> Support, support was what they gave
> They made me live.
>
> To die, not old, cocooned in love
> Transferring through a point in time
> To somewhere else
> A gift we have no right to ask
> And it is mine.
>
> My brainwaves span the cool night air
> And tap goodbye to friends unseen.
> Will they go on and span in time
> My after life
> I'd like to know.

Thank you all.

Sylvia, Julia, Heidi, Kieran, Nemone and Nicki.
January 1993

Contents

	Preface	19
	Introduction	23
1	Beaubourg	25
2	Early life	47
3	Sydney	59
4	Ove Arup the man	67
5	The role of the engineer	71
6	Jean Prouvé	81
7	Menil	87
8	Fabric	95
9	Glass and polycarbonate	107
10	Details: steel at Beaubourg, concrete at Lloyd's	115
11	Stone: Pavilion of the Future, Seville	119
12	The critic and the photograph	127
13	Working with industry	133
14	The Fiat experience	135
15	The chameleon factor	145
16	The Full-Moon Theatre	149
17	Architecture in movement	161
	Epilogue	165
	Buildings and projects: a chronology	167
	Bibliography	173
	Illustration credits	177
	Appendix 1: Peter Rice	179
	Appendix 2: RFR	183
	Index	187

Introduction

It turns out that con artists aren't much different from just plain artists. Like them they come in all sizes and shapes. Some are good at their job, others ineffectual, while some, like the fellow who introduced me to Peter Rice, operate in such a zany way that their motives remain obscure and unfathomable.

Responding to what now seems to have been a pretty far-fetched proposal I built a ten-foot-wide model of a footbridge – a spray-painted jumble of aluminium whose enlargement was meant to span the Seine. Since it matched his brightly coloured vision, my Parisian agent was pleased. He wanted to show it to France's minister of public works, but first he needed an engineering opinion. He asked Peter to look at my model. Surprisingly, Peter showed up.

'Well, what do you want to know?' he asked after walking around the maquette a few times. I thought, 'Oh boy, I'm really in over my head now', but managed to ask as casually as possible, 'Do you think it's buildable?' He looked at the model of the bridge again and responded 'Yes.' I didn't believe him for a second. Then I began to realize what that 'Yes' meant. Sure it was buildable – buildable by him, not me. But, fortunately, there was more to it than that. Somehow, even though he communicated a questioning, perhaps conditional, sense of approval he did it in such a way that the recollection makes me happy to this moment. It seems the 'Yes' implied that the model might be worth developing if we could work it through, if, first, I could only make myself clear about my idea for the bridge.

We never got very far on the bridge, but later we did get pretty far on a museum proposal for the Dutch town of Groningen. On that project I got lucky. A simple, communicable idea popped out at me from a Dover book of Chinese lattice designs. Twisting one of its leaf shapes made a wonderful roof plane for our building model. When Peter asked me what was the idea behind the wavy roof I could say proudly, 'It's like a leaf.' Once he had a handle, once he could grasp the image, Peter just rolled on like a juggernaut, crushing the obstacles of practicality and cost, making it possible for us to build what we liked.

I guess it's obvious that Peter is a national treasure but I feel the truth is that he is a universal treasure. Just to be around him makes you want to think, think as hard as you can. Even when the results are worthless and the efforts futile Peter has helped me realize what a privilege it is to be able to think at all.

Frank Stella, California, USA 1992

Aerial view of the
Pompidou Centre, Paris.

Beaubourg

Beaubourg, or the Centre Pompidou, as it is now called, is for me the beginning of my real career. In early 1971 I had been working in the Structures 3 group at Ove Arup & Partners in London, where I had been since returning from Sydney and the USA in the spring of 1968. The work in Structures 3 was a mixture of special cable and fabric structures, mostly with Frei Otto in Germany, and conventional building work in the UK. The Centre Beaubourg competition seemed an opportunity to experiment. The competition was run by the French government and the project seemed to us real and likely to be built. It was the cartesian clarity of the French administrative approach which persuaded us that here at last was a competition where one could expect everything to work properly. Of course today, having lived through the experience, we wouldn't be so naive, but then we believed; looking back, we were right to believe.

We, that is Structures 3 at Ove Arup & Partners – a group run by Ted Happold, where I was an associate – had been introduced to the architect Richard Rogers by Frei Otto. Ted had been approached by Richard about a competition for a new stand for Chelsea football ground. Frei had told Richard that he always worked with the engineers from Structures 3 at Ove Arup & Partners in London and together we made the submission, which was not successful, for the Chelsea football ground stand.

As we sat thinking we realized that a good reason for entering competitions is not to win them but to explore relationships and design. Of course one can hope to win but, particularly when it is an open competition – there were 687 entries for Beaubourg – to set out to win is in a sense self-defeating, because it will induce a conservative and tentative approach and the principal factor will be not to offend. With that in mind we approached Richard Rogers, who had just set up a joint partnership with Renzo Piano, and asked them if they would be willing to enter the competition with us. After some hesitation and indecision they decided to proceed.

The competition itself was a strange affair. Piano & Rogers had a clear idea of the building or image they wanted to explore – an idea stemming from Archigram, Cedric Price and Joan Littlewood, and the optimism of the 1960s. It was a large loose-fit frame where anything could happen. An information machine. At its core was the belief, which had been identified in the brief, that culture should not be élitist, that culture should be like any other form of information: open to all in a friendly, classless environment. Once the

An Engineer Imagines

'Interchange' by Archigram, the avant-garde group loosely based at the Architectural Association in the 1960s.

Front elevation of the Pompidou Centre.

Peter Rice, Renzo Piano and Richard Rogers celebrating Renzo Piano's RIBA Gold Medal Award, 1989.

architectural idea, the large open steel framework, had started to gel, our job, in one page, was to design the framework.

At first we sought literal solutions, large-span beams, giving clear space without columns. The result was lumpy, a frame which did not express our intentions at all. We decided instead to present an approach, one which would convey the spirit of our intentions even though the actual solution itself might be somewhat unreal. The structural frame, with large-span clear floors, was a simple problem structurally. It was easy to see how it could be realized. Therefore taking a different approach did not undermine belief in the project. Any engineer could look at it and say, 'not very realistic, this is what they should be doing,' and demonstrate the feasible solution.

By emphasizing instead a range of unusual ideas which fitted the philosophy and intentions of the architecture, we wanted to fire the imagination. After all, we did not expect to win. Piano & Rogers were doing the Burrell competition. That was the one they wanted to win; we were working with the architect Robin Spence on the extension to the Houses of Parliament. That was serious. But Beaubourg was for exploring ideas, and to see whether there was an interest in working together between us at Structures 3 and Piano & Rogers. No compromise was needed. We did what we liked.

The solution was based on moveable floors. That produced a detailing challenge. The joint supporting the floor became the central element of the design. We devised a theoretical approach with a water-filled structure to give fire protection. Making the joint the essence of the solution expressed concisely the spirit we wanted to convey. Paris, after all, has many marvellous steel structures, from the Art Nouveau entrances of metro stations, to the great structures such as the Gare de Lyon, the Eiffel Tower or the Grand Palais. Often it is the expressiveness of the jointing which humanizes the structures and gives them their friendly feel. We were part of a noble tradition and said so.

Doing the competition was fun. It was all done quickly near the end, so there wasn't any time for the fun to get lost. This is an important point about competitions, especially open ones. The entry must not become too deliberate or too detailed, or too closely argued a response to the brief, because the jury will only have the briefest of time with each entry. It is the idea they will see and the spirit of the drawings.

It is history now, but we won. I know the date, 13 July 1971,

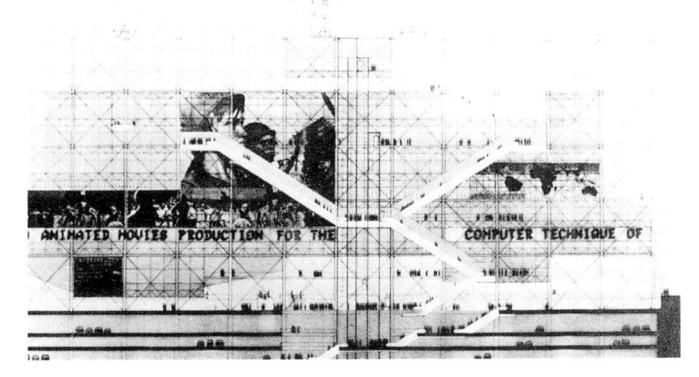

Beaubourg competition project elevation.

Beaubourg competition model.

Beaubourg competition drawing of structural system.

because my wife was in hospital having given birth to our fourth child, Nemone, early the same morning. Thereafter everything changed. The change was slow at first, but eventually everything changed completely.

I am not sure why one so resists change, with all the possibilities it offers. In the days after the announcement everything was confused. We, the engineers – Ted Happold and I – were not the main stars: that was the role of the architects. But we had our role, we were part of the team, accepted, necessary, stars to be perhaps. It was very exciting. It was I suppose like winning a lottery, which in a way was what happened. Everyone wanted a part. Within Arups everyone was prepared to help. Someone would turn up in Paris just because they thought Ted or I had winked at them in the street on Tuesday.

It took some time for reality to penetrate, and when it did it was daunting. In those first three months we had talked ourselves into a real role as engineers for the project. We had built Sydney Opera House after all, I was the living proof, and now we had to deliver. This was not how the French government had planned it should be. They foresaw French engineers and architects as the executive designers with the competition winners around to ensure that the intentions of the design were maintained. We argued that in a design like ours, it was all in the detailing. It was impossible to separate execution and intention. Robert Bordaz, the president of the client body, was convinced, demonstrating for the first (but not the last) time his wisdom and maturity in the face of many pressures.

The negotiations with the client and the authorities, with Ted Happold representing Arups, were complex and hard, and I did not have much part in them. I was slowly realizing the scale of the task we were so blithely insisting we could do.

I did not speak French then. As I am Irish, my less than comprehensive education had included Gaelic, Latin and ancient Greek, none of which was of much use in Paris in 1971. So we learned to smile instead. Perhaps it was just as well we did not speak the language. It is an advantage I have also noticed since, because we missed out on all the lectures and explanations directed at us by the great and the good. We just smiled at whatever people said and got on with the job of fulfilling our destiny. From time to time members of our own team who spoke French would interrupt the dream, and tell us how sceptical and unhappy the French officials were,

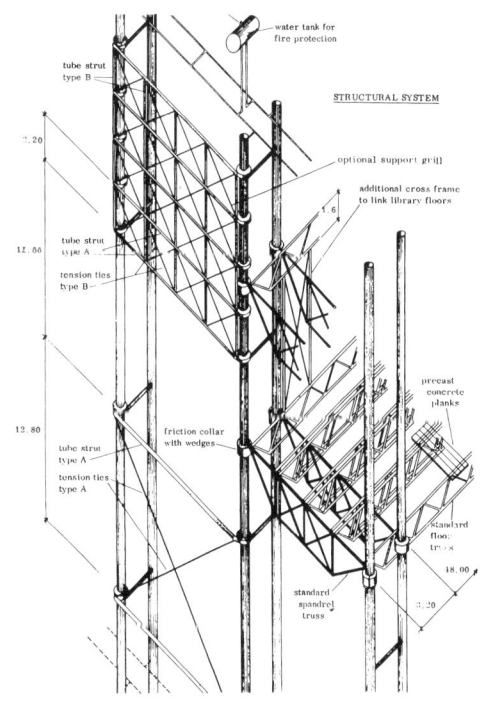

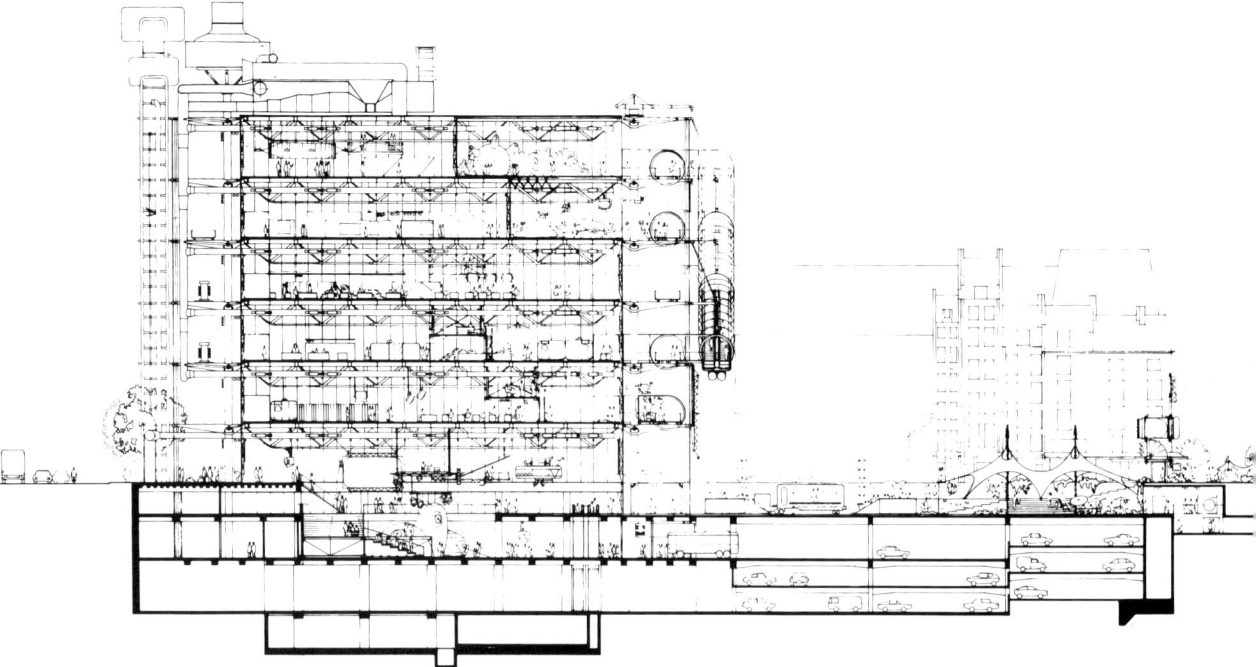

Pompidou Centre cross-section.

how unreal they considered our presumption, particularly the British engineers' presumption: 'We have some great engineers in France you know ...' But we continued to smile and believe that somehow we were different.

As a condition of the competition, the team itself was installed in Paris with both engineers and architects. It was small, introspective, isolated and also wonderfully *ad hoc*. None of us, with the exception of some members of the Piano & Rogers team, had really known each other before the adventure began. This meant that in the beginning there was some jockeying for position between the different team members, until a working relationship developed, all of which added to the air of unreality which permeated everything.

My own role in the project, and the relationship I built up with others, was dominated by the experience of Sydney Opera House. I had worked on that project for almost seven years. It was in its way the precursor of Beaubourg, a large international competition won out of the blue by an almost unknown architect. That too was a monument to culture, a symbol for a city. It was there that I had learned what little I knew about architecture. I did not know its architect, Jørn Utzon, very well, but I would follow him around site and listen to him reasoning and explaining why he had made certain decisions. The dominant memory was of the importance of detail in determining scale, in deciding the way we see buildings. Sydney Opera House was a big building, designed to receive the public; it was carefully detailed, down to the joint between the tiles in the tile lids. This had been Utzon's way of making the building soft and friendly. Beaubourg too was big, enormous really. It was going to require the same treatment, and the initial design decisions had to make that possible.

The essential engineering problem was becoming clear: a large, 44.8-metre span, carrying a heavy library, which could occur anywhere in the building. Future flexibility was a key philosophical

Spaceframe structure for the 1970 Osaka World Fair, showing the cast-steel nodes.

component of the design. A height limit of 28 metres defined the level of the highest floor – 28 metres is the maximum escape height of a firefighter's ladder. This latter factor was to avoid the building being classified as a 'tall building' by the fire authorities, as this would have meant many extra requirements for escape and fire protection undermining the whole concept. The other important requirement was that the structure had to extend from the front to the back of the building and this included the two movement zones at the front and the back. On the piazza side, the front, it was for the movement of people, and on the rue de Renard side this was for the services: the air ducts, cabling, plumbing, service lifts and all the other paraphernalia that a modern building needs.

Shortly after we had won the competition I made a trip to Japan to deliver a paper at a conference on tension structures. As part of that conference, trips were organized to Osaka to see the surviving structures of the 1970 World Fair. One of these structures was a giant space frame designed by Kenzo Tange and Professor Tsuboi, the eminent Japanese engineer. There I saw large cast-steel nodes.

An idea was born. I had been wondering for some time what it was that gave the large engineering structures of the nineteenth century their special appeal. It was not just their daring and confidence. That is present in many of today's great structural achievements, but they lack the warmth, the individuality and personality of their nineteenth-century counterparts. One element I had latched on to was the evidence of the attachment and care their designers and makers had lavished on them. Like Gothic cathedrals, they exude craft and individual choice. The cast-iron decorations and the cast joints give each of these structures a quality unique to their designer and maker, a reminder that they were made and conceived by people who had laboured and left their mark.

Cast steel could have this quality too. Cast steel had the possibility of achieving the same personal contact. But it had been abandoned as a building material since those same Victorian times. The need for reliability and predictability had effectively eliminated all those materials not produced by a standard industrial process where the absence of flaws was a proven outcome of the process itself. But here it was being used in a major structure. I decided Beaubourg would be designed using cast steel as its core material.

At this stage it is worth explaining more clearly the full logic which led to the choice of cast steel throughout the main structure

Cast-iron beams supporting brick vaults. Basement of warehouse at Albert Dock, Liverpool (now the Maritime Museum), 1848–51. Engineer Jesse Hartley.

Cast-iron aqueduct over disused railway near Linlithgow, Scotland (late 1830s). Designed and constructed by engineers of the Slammannan Railway.

of Beaubourg. Steel structures in buildings are usually made from standard steel sections, I-sections, channel sections, tubes and angles. These are produced by rolling and by extrusion in continuous lines which guarantee their quality. They also guarantee a visual and geometrical uniformity which leaves very little room for personal expression. In addition, everyone has a preconceived idea of steel structures made from factory-produced standard elements. Surprise and personality are missing, particularly for the general public, and with it all the feeling of contact and warmth between the person looking and the maker, whether designer or fabricator. The introduction of cast steel seemed a way to break this deadlock; a way to create out of the Centre Beaubourg structure something which, by its element of surprise and unpredictability, by the unusual conjunction of castings among the more common steel elements, would make the structure part of an approachable and sympathetic building. The scale of the Centre Beaubourg would be the scale of the pieces rather than the scale of the whole.

Of course there are many other more important facets to the building and its architecture. The movement of people on the piazza façade, the piazza itself, the large clear floor plans, are all part of the same intense effort to make the place a people place, where individuals can find themselves at one with this enormous monument. It was, after all, one of the express aims of the competition and of the solution offered by Piano & Rogers, to create a popular palace of culture, one in which ordinary people would not feel intimidated.

This sense of intimidation by culture was one with which I had a clear personal affinity. Coming from rural Ireland, where the only culture was a fœtid freedom of sound in language, by way of the technical departments of an austere Protestant university, places like the Louvre and the National Gallery in London overwhelmed me. The language of culture, alien and rigid and presumptive, was a barrier in itself. I really wanted to make the Centre Beaubourg an accessible place, a place, like the Sydney Opera House or the Eiffel Tower, where you could feel good, not afraid and over-humble.

Cast steel was the way to go. It became my contribution, a mission. I was, I felt, the most ordinary person there. I was determined to do my bit. Little did the bureaucracy know, or the industry which later on decided that this fixation with using cast steel was silly, out of place in the modern era of 1971, that they were dealing with an engineer obsessed. Engineers are sensible people, after all, not like architects or other artistic types. They are prone to make decisions by rational argument, as they are people with pragmatic values, people who have seen the light.

The decision to use cast steel was taken before we had a structure. It was made, too, before we knew whether we could make it work, before we knew anything really about how to design in cast steel. Three separate sets of work got going all together.

It is important at this stage to talk of teamwork. Teamwork is a much misused word, often used as a cover for blandness, or as a salute to others by someone claiming the credit and identity for some important artefact. Good teams are made up of different people, people whose separateness and attitude complement each other, and who by their individual willingness to work together and accept the presence and contribution of all of the others, for a while at least, make possible real momentum. Apart from Renzo Piano and Richard Rogers, the team working out the structure at this inception stage was made up of other key architects such as Laurie Abbott – another singular and dedicated nutcase, like me – a great draughtsman who could draw and make real the ideas I could only talk about, and three engineers: Lennart Grut, the man who controlled and did the work; Johnny Stanton, young, aged twenty plus, talented and at ease; and Andrew Dekany, an engineer of the old school, an engineer as engineers ought to be. Beyond and above all this in London were Ted Happold, Povl Ahm and the rest of the Arup hierarchy, negotiating and creating the space for the team to get to work.

The Arup back-up had a critical role. Working in Paris, we could have ideas and research and demonstrate how they worked. But, in engineering, structures must be correct. They must not fail. When innovating, which using cast steel in this way was, it is essential to have detailed and thorough analysis facilities available from highly skilled people with no emotional commitment to making the solutions work, just a clear, logical and objective insistence that the structure and its materials satisfy all the laws and requirements they should. Sydney had taught me how to use Arup and its gifted individuals. Beaubourg was the proof that it worked.

This use of the talent of the many highly skilled Arup engineers has been central to everything I have done, and my work would have been much poorer and much less adventurous without it. Two exceptional individuals, among many others I have relied upon

Lennart Grut, Ted Happold and Peter Rice, part of the Beaubourg team.

An Engineer Imagines

Gerber bridge over the Main at Hassfurt, 1866–67. The oldest metal cantilever bridge in the world, made up of three spans of 87, 140 and 87 feet.

Structural principle of a Gerber bridge.
A Bridge of single spans: easy to erect, but uneconomic structural depth.
B Continuous beam at bridge: difficult to erect, but economic structural depth.
C Gerber's solution: easy to erect and economic structural depth.

Double boom beam connects to gerberette, which is pinned to column.

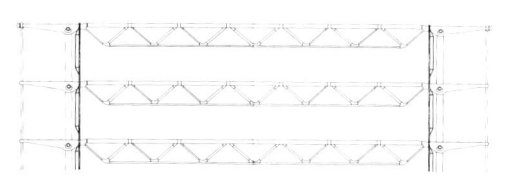

Cross-section diagram showing relationship between beam, gerberette and column.

throughout this long period, stand out: John Blanchard, the most gifted analyst I know, and Turlogh O'Brien, a materials scientist whose understanding of materials is second to none.

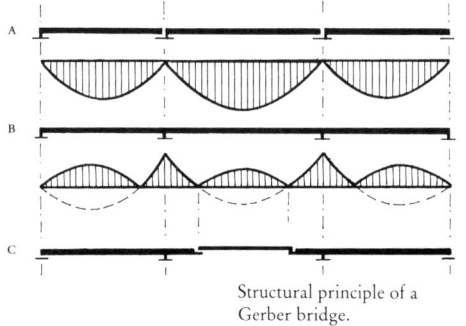

We quickly realized that the task ahead was not as simple as ringing up the Japanese and asking them the way. The clearest, if most intractable task was to find the right structure. The essence of the problem was that the architecture demanded a clear span of 44.8 metres, with movement zones on the front and back, which brought the whole structure up to 56.8 metres. The structure had to enclose all three zones. An additional complication was that the real internal space was only the main span – the movement zones were outside the façade – so a natural point had to be created where the façade would naturally go. It was no good making a fantastic solution, spanning the main space, with little outriggers to keep the moving people and ducts in. We tried of course to do it that way. It got no marks from the architects. We also tried double columns on either side, with the main beams spanning from half way between them. This got no marks either; indeed it might have registered a negative score.

Then one of the team, I am no longer sure who, probably Lennart Grut – I know it was not me – suggested a suspended beam on a short propped cantilever, the so-called gerberette solution named after Heinrich Gerber, a nineteenth-century German engineer who invented this structural system for bridges. This solution simply and elegantly resolved all the conflicts. Naturally it was quickly adopted. From here the next step was to design the elements – the column, the tie, the gerberette and the beam. The key to the whole was the gerberette. After all, a column is a column: a hollow round pole to carry load. And a tie is easily found; it became a solid round bar. No, the gerberette was the thing and then the beam. Again the struggle for a good idea ensued. A casting it had to be, but what kind, what shape should it take? The winning idea was supplied by Johnny Stanton, with help, if memory serves me right, from Laurie Abbott: a single piece, going from the tie past the column on either side to pick up the beam.

From here on the real design of the gerberette began. That took months. We had to make something that could be cast. Visits to experts, visits to foundries helped us understand how it should be done. The forces and loads in the piece – I like the word piece, it makes me feel like an artist when I use it – were the principal deter-

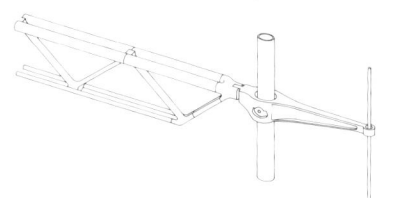

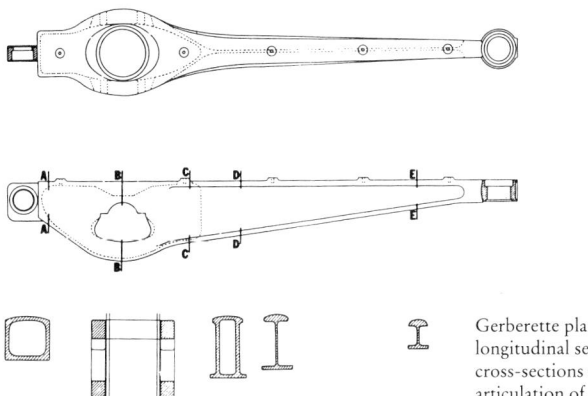

Gerberette plan, longitudinal section and cross-sections showing articulation of cast steel.

minants of its shape: slender at the tension tie end where the load is applied, deep and strong over the column where the load and moment reach a maximum, and slender again at the point of pick up of the beam. The development of form and the interactive nature of the design were complex at this stage. Certain decisions could only be taken by the engineers. The material quality and the nature of the stresses in the element had to be equated to the load cast steel could carry. Engineering responsibility in this situation is very clear. The structure must not fail. It must work, and continue to work under all conditions of load. It may look elegant or it may not, it may reflect the inherent nature of cast steel or the material of which it is made, and that is very satisfying if it is possible but if not you compromise a little. But it must carry the load.

Cast steel was a poorly understood material. It was principally produced by foundries dating from the mid-nineteenth century, especially in Europe in the heartland of the old industrial infrastructure. Their methods and approach were craft-based, and had not much changed since the end of the nineteenth century. To this tradition had to be added the requirements of modern testing for reliability and predictability. A new technology was emerging at this time, spawned by the need for reliable steel jackets for nuclear reactors and by the complexity of making oil platforms for the cold and deep North Sea. It was called 'fracture mechanics', the science of predicting the behaviour of metals under strain, and how they would react if small flaws and cracks existed within them. It seemed the answer. We worked with the welding institutes in both France and the UK to develop a way of transferring this technology to predicting the behaviour of the gerberette, and defining what the material had to be. The final shape of the gerberette, carefully drawn, modelled and redrawn by the architects many times, reflected all these needs.

Gradually, as we worked, a philosophy for what we were doing emerged. The piece began to absorb in a very visible way what was happening. Its shape was almost a structural diagram of the forces within it. The needs of casting were shown by the thicknesses of the top and bottom. The shapes of the openings at the top and sides were produced by the erection sequence. Large spherical bearings were introduced at the point of support of the beam and at the point where the gerberette sits on the axle through the column to make the column work properly. All these things are visible. It had

Workmen installing a spherical bearing.

Construction shot of assembly.

become an essay on how it works. All points of contact between the gerberette and other elements were machined. This had the unintended but very important advantage of controlling the way any future user could attach elements to the casting. Future flexibility could be controlled. The more you know, the better you can read and the more you can see. All the information is there, and the gerberette is consistent with the nature of cast steel as well as being a shape whose form is a direct result of the forces acting on it. And yet it is quite personal, to whom exactly I am not too sure, uniquely a product of that group of people working on its development, our team. And although manufactured by industry, it is not a hard impersonal industrial product but one where human intervention is very evident.

The essence of the design given by the use of cast steel was that each piece was separate, an articulated assembly where the members only touched at discrete points. As in music, where the space between the notes defines the quality, here it was the space between the pieces which defined the scale.

The design having been chosen, work proceeded to draw and define the whole of the steel structure so that it could be tendered. This work was done in an atmosphere of considerable scepticism. This scepticism came from the different French parts of the team, from the client and from all the different advisors, mainly French industry and expert advisors, though not the people from the French welding institute, the Institut de Soudure. Basically the sceptics could not understand our motivation, or the logic of our position. 'What do you want?' they said. 'You want it to look like that – OK, we can make it to look that way by traditional methods, just leave it to us.' Implied was: 'You're too young. The world is not that simple. Just let the people who know get on with it.' We, for our part, ignored all the talk. Even when we understood, we smiled and worked on.

The other elements of the design, the beam, the centrifugally cast columns and the ties which were radical in themselves, paled in comparison to the gerberette. The gerberette was the symbol of our daring, the core of our attention.

The groups that were to participate in the tender were assembled and the project proceeded as normal, or almost normal. The client and president of the project agency, in the person of Robert Bordaz, showed enormous *sangfroid* in the face of all the doubt and scepticism. He effectively protected us from the real criticism and never once asked us to stop. This was extraordinary considering the difference between the credentials of the protagonists: on the one hand all the senior figures of French industry who would have to build the project and be responsible thereafter, and on the other us, a group of young foreigners, almost without a track record. Admittedly we had won the competition to be allowed to design the Centre, but even here the scenario was not going exactly as the client had scripted it. In the original screenplay the client would hold the competition, a winner would be selected who, if from overseas, would be married to competent French professionals who would effectively execute it. And indeed, after tiring through the experience of Beaubourg the French government has ensured that in all subsequent competitions that is how it happens. Not only was there this scepticism in French industry, there was also considerable opposition from French professionals at the slight implied by the fact that they were excluded from the most important job to be built in France since the war. But the client resisted it all, and gave us the space and time to carry out our own design.

It is interesting to reflect on what being a foreigner means in this kind of situation. Not only did we not understand properly the level of disquiet opposite us, we were also unaware of the undercurrent of rumour and doubt prevalent in the society at the time, both about the project and about us. At home, all this would have been an ever-present factor and would undoubtedly have dampened our ardour and sapped our confidence, and maybe eventually forced us to change.

During the tender period we presented the project to Socotec, the *bureau de contrôle*, or checking authority. It was not impressed. Almost everything was wrong. Like a boxer getting up after a count of eight we steadied ourselves and tried to understand and itemize the comments. There were so many that even this was almost impossible. Many were complex and very negative and we concluded that the *bureau* had no belief that the project would be built. Unlike us, its members could hear the undercurrent of criticism. We proceeded nevertheless to tackle the items one by one, and with the help of Robert Bordaz we started to achieve a breakthrough. By the end the intervention of Socotec was very positive. M. Daussy, the chief engineer of Socotec, suggested we load-test all the principal elements to guarantee their integrity. This was without doubt the most intelligent solution and one which proved a godsend when fabrication started.

The tender itself proceeded without incident until the results arrived. Then all the non-French contractors withdrew and the two main French companies combined to offer a single price, at 50 per cent over budget, and also put forward an alternative structural solution which mysteriously cost exactly what we could afford. All hell broke loose. The client declared the tender null and void and instructed us to negotiate with the lowest bidder, while exhorting us to search for alternative companies worldwide who could build our scheme. They were sceptical and I think believed that a negotiated compromise would be the probable answer. GTM, the main contractor, suggested a concrete alternative, to indicate to the steel industry that it would not have it all its own way. And we got through contacts two positive responses, one from Nippon Steel in Japan and one from Krupp in Germany, who were upset that they had been effectively excluded the first time around.

The day the Nippon Steel telex arrived the air began to clear. Yes, it said, it was all possible, and it could be delivered free-on-board at Le Havre at five francs a kilo, a price exactly half our budget. Slowly belief returned. And we began to see real progress in our negotiations with the French contractors. Krupp too sent an acceptable offer. We were on our way.

I would like to say here that this situation was not just a problem of French industry. I have seen the same circumstance arise in many countries, and have resolved it in the same way. A recent ironic twist to this story arose in Japan in the late 1980s during the Kansai Airport terminal project, where the same Japanese firms which were so helpful during Beaubourg were resisting the solution we had proposed for the terminal roof structure and end-wall glazing as being 'impossible to do, not Japanese enough, not understanding …', and the same French firms which were so entrenched and difficult in 1973 became the saviours in 1991, undertaking to build exactly what was required within budget and on time. There is no monopoly on the beliefs held by industry, particularly well-entrenched industries like the steel construction industry. From Tokyo to Paris, New York to Seoul, the rule seems to be inflexible and opinionated at home, helpful and flexible overseas. Never believe what industry says is possible or impossible. It is all said with some other motive in mind.

We began to negotiate seriously with Krupp, to make sure it understood the scheme correctly and to get a clear and proper picture of its price. During this time French industry too began to realize that it might be advisable to look properly at a solution. It started to discover that this wasn't so impossible after all. Well, yes, maybe they could do it. And well, yes, perhaps the budget that we had was achievable. But we continued with Krupp, and the client too kept faith with Krupp.

In this and in many other things the French client was remarkable. It was a special agency called an *établissement public*, which has the characteristic of not depending on a single ministry, especially created to define, specify, construct and set up Beaubourg in all its aspects. Robert Bordaz, the president of this agency, was a very senior civil servant who had close contact with the real client, President Pompidou, and therefore had enormous freedom of decision to do and to carry through what he believed was right for the project. He and the *établissement public* were responsible for every aspect of the project, from the purchase and acquisition of works for the Museum of Modern Art, to the appointment of all the centre's directors and staff and the control of the budget. The definition of what was in the true interests of the project was ruled by him, informed by the opinions of advisors – including the leading members of the competition jury, notably Jean Prouvé, the celebrated French inventor and engineer, and Philip Johnson, the well-known American architect.

This support enabled us to have the steel structure as originally defined. In 1973 a contract was signed with Krupp, Krupp of Big Bertha fame, a name to strike fear and unease in the hearts of many older Parisians, but also a Krupp making manifest the progress to a united Europe; proof, if proof were needed, that France at least believed in a European future. At this point a new saga began.

Krupp, and Polig Heckel & Bleihart PHB, the foundry from Saarbrücken, got quickly to work. Drawings appeared and nothing of substance was changed. Within four months they were casting the first beam nodes, and within six months they were ready to cast the first gerberettes. It was exciting and satisfying. Our great œuvre was taking shape. We should have been warned. Things don't go that well normally; I am very cautious when things go too well. But what can you do; a positive atmosphere allays even the most suspicious mind. And then the blow struck in early 1974. The first gerberette

Cast gerberettes in foundry.

tested failed its proving test at half the design load. It broke in two. And then the second one failed as well, and then a beam met the same fate. Something was seriously wrong. Some forty-five of the gerberettes had been made by this time, together with enough beam nodes for half the beams, with six of them fully fabricated.

Time was tight. One of the conditions of the competition had been that the winning design should be made and in use in five years. In our entry we had concluded that our design would take maybe six years. But the jury, considering that the design was simple and clear, had decided that we were too pessimistic, and in retrospect I expect that they were right. What we had done was to extrapolate

Steelworker fettling gerberette in foundry.

UK construction practice to the French, but the latter had habitually built much faster. The five years were agreed and became an obligation of our contract. The implications of this were passed on in the tender conditions for the steel structure. But about four and a half months had been lost in the argument over the tender, and what was initially a tight contract became tighter still. Krupp accepted all this when signing. But what to do now? We went to Germany and started to investigate.

What we found was a classic case of international misunderstanding. Hindsight is a terrible thing, and here it was all so obvious. Our tender had been written in French and had used French

Transportation of beam to site at night.

national standards, as was right and necessary. They were the only ones which had legal status in France. On one point we had used a special British standard, then under development. This was the standard which specified the material quality for the castings and the special welds in the beam. It was a British standard because most of the work on this new approach in fracture mechanics had been stimulated by the problems of making the North Sea oil platforms, and the British Standards Institute had perforce to define a way of measuring steel quality in these conditions. Similar problems had not arisen in France or Germany at that time. So Socotec, and the other French authorities, such as the Institut de Soudure, had accepted this unusual approach. The British standards were cross-referenced to French standards at the tender stage so that companies unused to the esoteric requirements of this special British standard would have a general idea of the standards and type of steel required and could therefore price the material quality at the tender stage. The two principal German contractors on seeing this had found the equivalent German standard and, being assured that the German standard was more rigorous than the equivalent French standard, had proceeded to make the pieces to the German standard without reading or trying to understand the nature of the requirements implied by the new special British standards quoted. In fact these standards were quite different at the special level of the pieces we were making, especially the gerberettes. These had large thick sections which behaved in a quite different way from the thin sections treated by the normal standards of the day. Our German colleagues spoke little English, we spoke no German, and none of us was very good at French.

Communication was difficult. The essential problem was to find a way to correct as quickly as possible the pieces already made, and to find the best way to make the remainder. But first we had to find a way to communicate. We searched in the UK because at least there we were sure to understand each other. The French client and main contractor put their faith in French experts, and the German steelmakers, the ones with the biggest problem, who neither had faith in nor particularly trusted us or the French, looked in Germany. They just could not understand how an obscure British standard could be right and all the German DIN standards, the most rigorous and stringent, could be wrong. The situation was not easy.

At about this time, there was another event which brought home with instant clarity the seriousness of our situation. On 3 March 1974 a Turkish Airlines DC10, returning from Paris Orly to London, carrying many English rugby supporters who had been in Paris for the France versus England match, crashed killing all on board. It became a *cause célèbre*. Investigation, particularly in the English press led by the *Sunday Times* Insight team, soon established a catalogue of errors and lax control at the airport. A baggage door at the back of the plane had been inadequately secured and, under the extreme internal/external differential pressure at altitude, had burst open. The resulting explosive decompression had caused the passenger-cabin floor to collapse, severing or jamming the control cables which ran through the floor from the cockpit to the tail. Without a central hydraulic system, and thus no alternative means of steering the plane, the crash was inevitable. The investigation also showed that during the design someone had suggested that this might not be a good thing, but had been either ignored or overruled for political motives. It was clear that the designers were negligent, or at least that was the presumption of the jury of journalists. It had been pointed out to the designers that what they had done was wrong and they had ignored the advice. What could be clearer?

I began to ponder our own situation. Here, too, there were those who had said 'don't', more than just a few. Our tests had picked up the failure but a solution had to be found; otherwise we too would be subjected to trial by journalist. The thought of this spurred us on. Our search for a solution led us to Stuttgart University, where Professor Kussmaul at the Institute for Materials had used the fracture mechanics methods. His experience had come as part of the German nuclear programme. Suddenly Krupp and PHB had a German engineer telling them that our methods were right, the only way to go. A solution was soon suggested by a colleague of Professor Kussmaul. It saved the pieces already made by reheating them. Some had to be rejected because they had been machined and could not be heat-treated after machining, but the bulk could be reused. The beams were rescued in the same way. Although three months had been lost, a disaster had been averted and the client got Krupp to agree to a further shortening of the programme time. Shortly afterwards, the first pieces arrived in Paris and the project proceeded as though no crisis had ever existed.

Before this final relaxed phase, one more surprise was in store. In May 1974 Pompidou died. During the interregnum created by

Gerberette being lifted into position.

An Engineer Imagines

View through gerberette zone with shadow relief revealing the sculptural qualities of the gerberette made possible by the plasticity of the casting process.

Gerberettes punctuate column. Space between gerberette and column lightens and articulates connection.

the election of his successor, the Centre was renamed the Centre Georges Pompidou. Robert Bordaz canvassed the three leading candidates to ensure the best support possible for the future of the project. But, whoever won, the support would never be the same again. The eventual winner, Valéry Giscard d'Estaing, was probably the least supportive of the candidates of the right. He considered the expenditure a misuse of money and probably did not like the design. He was somewhat compromised in this by the fact that he had been minister of finance under Pompidou. This did not inhibit him. He demanded that floors be removed from the building. Luckily we had gone too far for this to be done.

It was summer when it happened. I was on holiday, camping in Normandy. I had been very careful not to tell people where I was going. And I had not planned the trip. I had collected the car with my family at Cherbourg and had scouted out a suitable camp site. Nevertheless I was astonished to find one morning a man chasing me along the beach, shouting, 'You must ring your office immediately; you must come back to Paris.' Naturally I assumed the worst. Something had failed. I rushed to the telephone. It was lunchtime, no-one was in. Even in France, I thought, real disaster would take precedence over lunch, but then maybe not. I could not be sure. I spent an anxious hour until finally I got through. 'The President wants to take two floors off the building. You must return.' 'Is that all?' I said. I was so relieved, I readily agreed to return; not a decision that was met with much enthusiasm from my family. And it was too late for change. The design had survived another attempted coup. From there on the project proceeded, with no more than the usual bumpiness, to open as planned on 30 January 1976.

Other elements of the design had been completed after the gerberette saga. But these were easy. When designing the main steel elements we had created a language of design. We had tubes in compression, solid rods in tension and cast elements for joints. Once these rules had become established everything else was easy. And no-one, no contractor, no expert doubted our resolve, our ability to have the solution we wanted. Dialogue was simple. People listened, and we even spoke some French. One day, we thought, we will be accepted.

Sub-teams arose within the main team of architects and engineers in order to design and realize other elements. The façade was under the control of Eric Holt and four other architects; Michael Davies

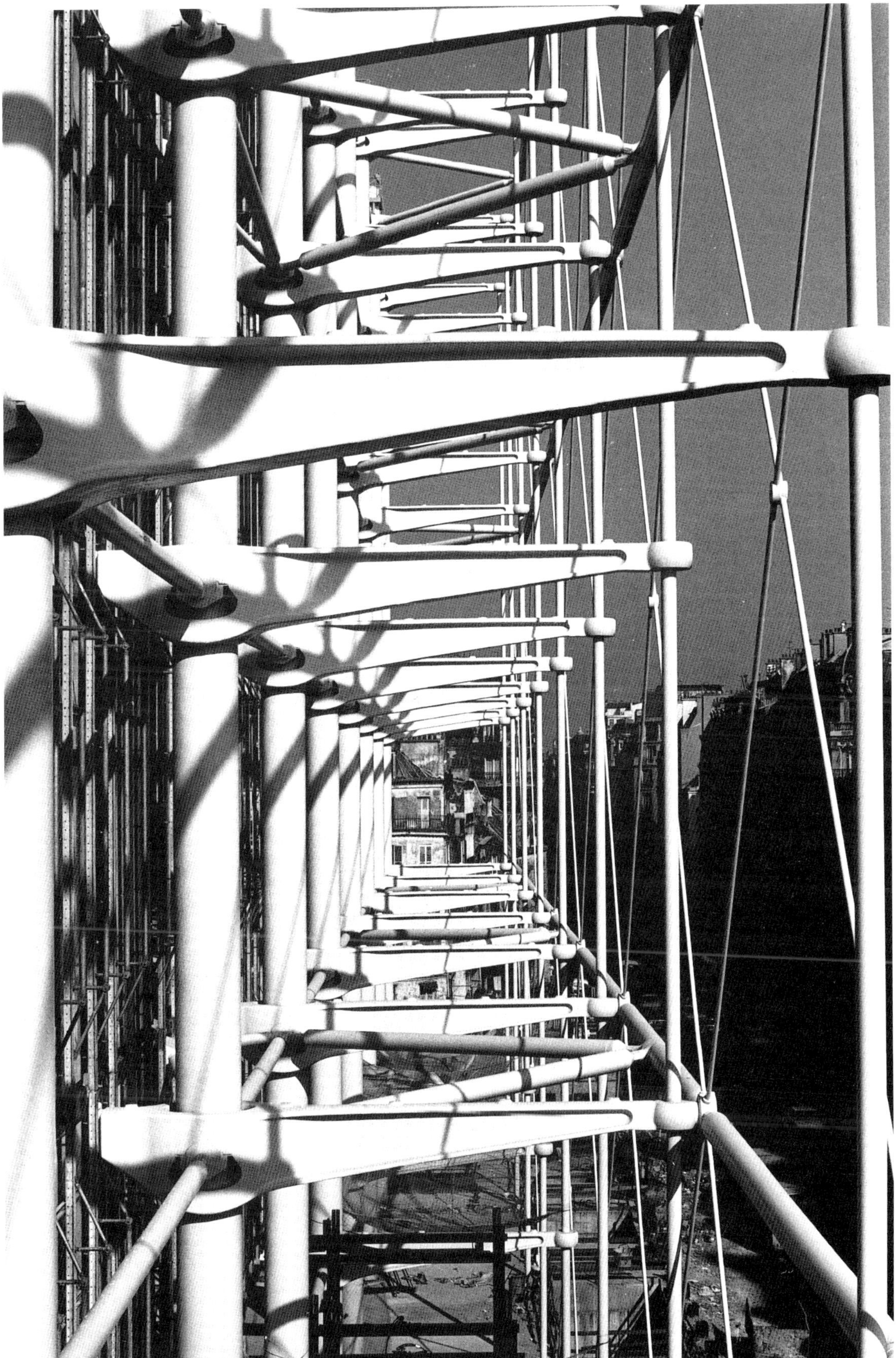

designed IRCAM (the Institute for Music and Acoustic Research) with yet another team; and the piazza and its surrounds became the œuvre of Cuno Brullmann and Alan Stanton. Gianni Francini and John Young were two more team leaders with their own teams tackling different sections. Walter Zbinden was in charge of the substructure. In another team, my engineering colleague Tom Barker developed the servicing strategy and details with Laurie Abbott. Separately the solutions adopted reflected the input and leadership of each individual, and although everything was approved by the real architects, the ideas often came from below. The final acknowledgement that should be made is the role of GTM and especially of Jean Thaury, the site manager. He too had much to bear but was positive and helpful throughout.

The essential lessons of this experience are simple. Communication is the key to progress. But communication is very difficult. It requires two: one who communicates and one who listens. It is impossible to measure or to know whether it has been successful. Communication with industry is especially difficult. Industrialists live in a definitive world, one which they control and are used to controlling. Everything is heard through a screen that filters out all unnecessary information. Any information which might threaten the equilibrium becomes distorted and misunderstood so that, in the end, one can only work within the confines of an already existing technological environment. Our attempt to marry the modern approach of fracture control with the traditional craft industry of cast metal was, in a way, doomed to fail initially. Like the Berliners of that time, we were shouting across an impenetrable wall, with each of us drawing the conclusion we wanted from the discussion. We needed our contractors to understand, so I suppose we ignored any sign that they did not comprehend the magnitude of what we asked. They, for their part, did not even read all the paperwork we had so laboriously prepared.

We were, I think, seduced by the legal protection the whole system seemed to offer. After all, we were right. It was written down. Krupp and PHB had signed a contract. Everything was clear, nothing was left to chance, the contractors knew what they had to do. Our drawings were clear and concise, even they admitted that. But what if they could not do it? What if we could not find a solution? Legally our case would be watertight, but what good was that? Without a solution we were as vulnerable as the contractor. In our modern society, as it becomes ever more legalistic, we can become dazzled and unable to see the simple needs, the simple objectives. The jargon of contracts is fine for lawyers and for the settlement of disputes but it can grievously corrupt real dialogue and communication. This is especially so when more than one language is involved. It is then that one realizes that language is culture. A Frenchman reading our French contract would not have made the same mistake as the German contractors. He might have made others, but they would at least have been consistent with some of our expectations. Translation does not solve the problem. Hierarchy and relationship remain distorted. And this is impossible to bridge by translation. It was not until we found a respected German expert who explained the problem that any recognition showed.

All those long meetings spent cajoling and carefully explaining the real issues were like the ozone layer: a problem for someone else. The contractor knew what he was doing. This was his *métier*. He had been doing the same thing since forever. There was obviously a simple explanation. He would do it better. He would watch the process more closely. How could he be wrong? He was after all the *maître* of his technology. To be lectured by young foreign designers on how to make the basic artefacts of his *métier* was unreal and probably insulting. The contractor might have been sweating during our sessions but it was not the sweat of concentration, it was the sweat of worry. It could not be true. As for the specification, what good was that? Architects and engineers were constantly writing harder and harder specifications. These were only designed to make life difficult, to tie the contractor well and truly in knots. Thinking that the specifications might contain the solution was not only unlikely, it was being stupid and unhelpful.

And so communication and dialogue were inhibited wherever we turned, by convention, by prejudice and by fear of failure. It was also the problem in the beginning when we embarked on our design. We failed completely to explain and convince. It needed a series of traumatic events to make even simple language understood. In a way the only communication that worked well was between architect and engineer – a dialogue which quite often fails. This is a tribute not only to the architects' willingness to enter discussion, but to their willingness to give their head to those around them, to trust others with decisions which ultimately would become their responsibility too. I suppose it comes of confidence

Renzo Piano, Robert Bordaz and Richard Rogers.

The 'piece' with the people.

in oneself. Whatever it is, it is not often found.

And the client. What can be said of him? Rare is hardly the word. *Sage*, as they say in French, and mature. One may like or dislike Centre Pompidou; one can certainly find many who dislike it, and its raw angular vitality would certainly be tempered if designed by the same team today. But I hope it can be accepted that it was worth doing and worth doing correctly. It was almost a necessary step, an acceptance of the reality of the technological world we live in and how it can be tamed. That is, I hope, what the engineering story is about: the attempt to make technology personal and identifiably human. Without the client, and M. Bordaz in particular, none of this would have been possible.

Was it worth it? No doubt at all. For me that question is simply answered. One day shortly after the opening I saw an old lady, dressed in black like the Irish mothers of my youth, sitting perusing the people and looking in wide-eyed astonishment. I watched her for a while, just sitting quietly, stroking the side of the gerberette, she was not afraid, not intimidated and she was on the fourth floor. And I thought that if somehow, by introducing elements like that, we can make people – people who would normally be alienated by things – feel comfortable, it proved to me that the thing that really matters is to introduce elements and materials into buildings in a way which reflects their real nature. The effect might not all be due to the cast steel, but that was certainly not negative. It was a positive contribution to the atmosphere and ambience. It was worth it.

Early Life

Inniskeen Road: July Evening

The bicycles go by in twos and threes –
There's a dance in Billy Brennan's barn tonight,
And there's the half-talk code of mysteries
And the wink-and-elbow language of delight.
Half-past eight and there is not a spot
Upon a mile of road, no shadow thrown
That might turn out a man or woman, not
A footfall tapping secrecies of stone.

I have what every poet hates in spite
Of all the solemn talk of contemplation.
Oh, Alexander Selkirk knew the plight
Of being king and government and nation.
A road, a mile of kingdom, I am king
Of banks and stones and every blooming thing.

Patrick Kavanagh
from *The Collected Poems* (London, 1964)

Dundalk, Gyles Quay and Inniskeen are separated by no more than 15 miles, Gyles Quay to the east of Dundalk, Inniskeen to the west. In 1946, when my grandfather Dada Quinn, who was schoolmaster at Inniskeen National School, died, they buried him in the village cemetery, to lie we thought forever in the womb of the village, surrounded by those who admired and respected his word.

Even to an eight-year old, Inniskeen was a claustrophobic place, small fields, tall hedges and whins, with eyes following every move. Outside was more enclosed than inside the big house. Maybe it was the colour; the high, dark, brooding green hedges, threatening to engulf you for straying in their midst. As a visitor there was no freedom to roam, the fields belonged to others. You could go where you liked but you walked with fear, a permanent heavy shadow covering everything. The house had one big field, with a small orchard and a stream with an otter that I never seemed to see.

And the border, three or four miles away, it was near, and talked of all the time. It was how people lived, the border with its mysterious Protestants and Royal Ulster Constabulary beyond. Fearful is not a good way to be in the country; you get no respect, no right to be young and inquisitive. It was not the fear of what was there but of darkness, which came suddenly with the rain, and of lightning, which if it caught you would dance between the whin bushes on the hill to test how you ran. It was fast and menacing, and made everyone afraid.

Dada Quinn was gentle but old, remote and unreal. Too many children had passed his way. I was his favourite, they said, and I do remember him smiling and holding my hand and testing me on mental arithmetic; a healthy man, always walking, going for walks, you had to talk because then he smiled. My mother was proud that I was his favourite. I always got the answers right to his questions.

I knew I loved numbers. They all seemed different, with their own special quality. Each number was precise, it meant something. I could not read. It did not seem to interest me, but with numbers I could play all day in my mind. Not like words, which flowed endlessly overhead. I was happy with numbers and Dada Quinn in his stern way was the only one who seemed to understand that.

Numbers and gooseberries. The taste of gooseberries is with me still, ripe and soft and cracked, sometimes hairy, sometimes smooth, and occasionally with a red flash, which tasted just the best. There were blackcurrants as well, and early apples, Beauty of Bath I expect. We went to Inniskeen in July, but the gooseberries were the reason to be there.

One summer there at the gate was the man who dug the ditch, Patrick Kavanagh. I was told he wrote poetry and had no money. He had been to the village school and was really above himself, but he was nice and could understand about being afraid. But then he was gone. Of course it was later, much later, that I realized that he was a very fine poet, whom I would have loved to have known.

The prejudice of the country was total. Everyone had their place and to think above your station was not acceptable. Inniskeen then, as today, was not an open place. It housed many dark secrets, and an inquisitive youngster from the town was not welcome. The family connections helped and one great-uncle, who part-owned the mill and two sizable farms, had a big and friendly house.

Haymaking on a warm summer's day was the opposite to the dark whin-covered hillsides. You could ride the dray at haymaking time and slide down the haystack. The sons, my mother's cousins, were local heroes. Three of them played Gaelic football for County Monaghan, and two even played for Ulster. They lived in another world, a world of distant journeys. Gaelic football was not for the physically timid and, although I was well co-ordinated and might have played, I never mastered the art of jumping and catching with abandon. I thought when I should have acted, and worried. I was too much a dreamer for this essentially country game.

You got to Inniskeen by train, a train which went from Dundalk to Bundoran. Bundoran, Bundoran, how romantic and far away it sounded. Sometimes my father would go to Bundoran when he wanted a holiday on his own, to play golf and be one of the lads. But we never went beyond Inniskeen, not even to Carrickmacross, five miles away, and certainly not to Castleblayney or Bundoran, places whose names even today have a romantic resonance born of watching the train pull out of Inniskeen station and head west. Later, after the war, we had a car, an Austin 8 – mothballed in grease for all the war years, to emerge tentatively into the post-war light.

Although Inniskeen is one of the core memories of my childhood and Gyles Quay is the other, we actually lived in Dundalk. But Dundalk was not a happy town. It was a hard town, an Irish industrial town in a country where there was not much work. My father was the chief education officer, which meant he was the head of technical education for County Louth, the smallest, but one of the most populous of the twenty-six counties.

It was a job of great influence because many of the local jobs got filled on his recommendation. In the country especially he was constantly wooed by small farmers to get their sons jobs. This was just as well, because he was an anglophile in a Republican town, still smarting after the Black and Tans and the civil war.

My father was not from Dundalk but from Kilkenny, a town so far away that in Dundalk he was an outsider, without a family history to define him, and he kept his local contacts to occasional visits to the bar with a few non-political friends. He was a gregarious person, a great chat, when he wanted to be, but mostly he kept to himself, taking long walks and listening to the BBC.

He had studied for two years at the London School of Economics in about 1920. He had been selected with fifteen others to go to university in the UK by a suddenly panicking administration, who could see home rule or independence on the horizon with no-one trained to run the country.

He was at the LSE at a time when the great names of socialism held the school in their grasp. From this he developed a life-long agnostic socialism, almost communist at first, which was definitely out in the Ireland of the 1940s.

Indeed he had to be extremely careful that his wayward political enthusiasms did not show because, although appointed in Dublin, he was responsible to a local committee whose dominant figure was the parish priest. Twelve o'clock Mass every Sunday, at the back of the cathedral, was the necessary minimum – with the other renegades, congregated in a visible place, men who had little in common except a certain difficulty with priestly authority. Father Stokes himself, the parish priest, understood that not everyone was destined to be devout; I think he preferred it that way, as long as you knew the rules. As often happens, an agnostic middle age gave way to a devout old age for my father in Canada, a place where you could please yourself.

In Dundalk I went to school at the Christian Brothers, the centre of rigorous Irishness, where everything was taught in Gaelic. This was following my father's theory that if these were the tenets I had to live by, I had better learn them early and understand them. It was a tough school where beating was the order of the day and things were arranged that at least once a week – and usually twice – you fell foul of authority, especially if, like me, you were erratic, always seemed capable of doing better, and not from a good family. A good family was one with an exemplary record during the troubles. I did not mind the beating; it was controlled, a leather strap on the hand. Sometimes you had to cry, otherwise they would not stop, but mostly it was a stoic acceptance of what was an inevitable part of life. Being beaten was infinitely preferable to being singled out or being teacher's pet. The classes were large, forty-five or more pupils, and up to four years in age difference. As I was one of the youngest I had to be careful not to show off, not to antagonize the bigger boys, who were hard and unsentimental, some of whom later became founding members of the Provisional IRA.

This rigorous republican schooling, which was not so bad however it may sound, was modified by a totally different attitude at home. There my father supported the British in the war and loved to listen to Churchill, the archetypal English ruler who had dominated Ireland for so long. This education was steeped in Irish views, by writers and teachers who were finding their way to an Irish identity, not an easy thing in a small country where the real language was English. It had a strange by-product.

History, social values – that is religion – and literature were selected to tell the story the Irish way; English history taught as part of European history was given the same cursory more-or-less objective treatment as that of France, Germany or the Papal States. The great heroes of England's past were reduced to size, and their warts were not ignored, so all the re-evaluations of the last thirty years have just about caught up with what we were taught anyway. It was so nice to watch television in the 1960s and gloat as Elizabeth I became the ogre she had always been to an Irish sceptic.

Mostly I did not fall foul of this dichotomy in my life, although my father was hardly kosher and my mother's family had been on the wrong side in the troubles, for Michael Collins in an avidly Fianna Fail area. Occasionally things went wrong. On one not to be forgotten day I was paraded as a trophy from the cathedral to a field near my home, with the promise of a beating up at the end. It happened after a boy's novena, which always took place on a Friday. The reason was that my captors were upset with my father. We lived in the end house of a terrace on the corner, with a big gable wall which was plastered smooth to waterproof it. It was ideal for handball and was in constant use for this purpose by the bigger boys. This one thing would drive my father wild. He could hear the sound and he blamed the handball for the leaks that came through the wall. It was a continuous war. That one week my father had threatened to call the Garda. The boys were intimidated, so they decided to teach me a lesson. I was picked up on the Friday evening and paraded all the way home. All my father's sins came out and I was properly frightened, but in the end there was no beating and, apart from a fast run home, no outcome at all.

Growing up in this Ireland of forty or more years ago there was no stimulus to think. The great events of the world were far, far away and did not seem to touch us. I knew nothing of the war, which seemed to be happening in a far distant place. Once or twice a plane would pass over and drop bombs, in the mistaken belief that Dundalk was somewhere else. The bombs did no damage and we soon returned to a familiar pattern of life.

Freedom was physical – to run, to swim with energy flowing – not intellectual. Maybe I was not very inquisitive. Thought was circumscribed by sin. 'Free thinker' had the same malevolent sound as 'racist' today. How could you think freely when great scholars had sought guidance for hard-won clarity all the way back to St

Augustine? Classical clarity was what we sought. Freedom, if it existed, was in mathematics where great ideas could be explored without boundary. I remember well a period of intense interest in – and great envy of – the achievement of Galois, a young nineteenth-century French mathematician who discovered group theory. I struggled to understand it, without much success, and fantasized of the freedom to roam in this elegant and precise world.

Fantasy and physical freedom were the points of escape from conventional belief. There were also of course the great traditional Catholic fantasies, of St Teresa and Oliver Plunkett – an eighteenth-century Irish Catholic martyr – of St Francis and Thomas Aquinas, no Abelard for me then, that was much later. But these were difficult fantasies to sustain; ascetic and somewhat boring, they were far in the future, an ideal life that one day might be mine.

The real fantasies were physical and, as I grew older, sexual. They were of achievement and fame, and of a life in the world of far away. Strangely, I never did much to help make real any of these sporting fantasies. I was not good at team sports, inhibited by a lack of abandon – courage you might say – where physical contact was involved. Even in running or swimming, or the other solo activities I might have tried, I had no compulsive desire to become good. Of course there was not much opportunity but, even so, I never trained, never joined a club.

I was in a way a compulsive non-joiner: no boy scouts; I had no interest in golf or tennis, because to play you joined the club; no FCA (the military cadets). I was almost the only person I knew who did not join at least one organization. This strangely eclectic attitude, not disapproved of by my father, and not understood by my mother, left a lot of time for independent day-dreaming in my world of fantasy.

During this period there was one bizarre arrangement which, Gothic in its implications, left a profound memory. Each autumn shortly before Christmas we were taken to Dublin to see the pantomimes. When we went there we stayed in the house of my two spinster aunts, Nell and Moll, in a modern suburb called Mount Merrion to the south of the city. These two aunts, my father's sisters, lived with their brother, my Uncle Frank. Earlier in his life Uncle Frank had fought in Egypt and Palestine during the First World War and had returned with some trophies, strictly illegal of course. The principal memento was the unwrapped mummified hand and arm of a young Egyptian princess, which was kept in a glass display case on top of the piano in the front room. This hand, with gold-painted nails, was an ever-present reminder of the proximity of death. Occasionally I would be given a bed in the same front room, with appeals to bravery and male toughness which meant nothing of course, the dread and fear being all in the mind. Although the memory is sharp and vivid, I forgot all about it when I left Trees Road.

These two aunts were themselves quite bizarre: they were thin and tall with dyed hair and a fastidious sense of manners and habit in everything they did. They lived frugally and would save up to go on pilgrimage to exotic places like the Holy Land or Lourdes where you could win indulgences, credit in the afterlife in the 'Catholic Bank of Good Works'. This frugality meant that a hungry meat-eating twelve-year old was rarely satisfied by the fare they provided, even when it was pure and unadorned.

Of course I ultimately learned to read, at about the age of nine or ten, and eventually I started to read great books. I never read with abandon, maybe six or seven books a year. I read slowly; each book would fill my head, become absorbed and change me a little. *The Red and the Black*, *Madame Bovary*, *My Confessions* were three which resonate still. The books were usually French or Russian. I suppose I read to watch the heroes dodge and manoeuvre their way through life. Information never interested me. I can't read and absorb it. I learned the minimum necessary. It is still the way.

But I have been lucky. In our information-orientated society where everything is in reference to something, most people are in awe of mathematics. They feign ignorance, which is surprising, odd perhaps, but well understandable. So dialogue continues without threat – except from critics, those who decide what we think. Then I can feel diminished, small, uncultured. As indeed I should. With great work about which I know nothing, so many great ideas, everything so clearly connected, ignorance is a kind of arrogance. But that's how it is. And I think it is true for most engineers. They get lost in this world of visual culture, but not in music. Many are passionate musicians, not lovers of literature or history, and they fail to communicate in an intimidating world.

Thinking, which is what I think I do, comes freely. As a child I would imagine everything, all my thinking would become imagines. Every problem would get analysed by imagining this or that way out. All the time imagining things. It's still like that. My head is like a 1950s American breakers' yard, piled high with the trash of past imagines. I sometimes go exploring and discover them, and that can be great fun.

There was freedom to wander too. Most Saturdays we went by car to Ravensdale, newly planted with conifers on the side of the slieve. Here we (my brother came too) could roam and create a private world, with our private paths, mostly sheep or deer tracks, where we could run down the mountainside and no-one could find us. We dammed the stream and made pools of clear cold water. In the dark interior of the woods occasionally a sheep would appear from nowhere, bleating and sounding aggressive, but soon we would come across a clearing or a place that we knew and all would be well. It was immensely large and was never all explored, so there was usually at least one project to discover a new area where we had not been before. Then we would have to run home, to be there at the entrance before my father had finished his long walk, and it got dark, and the whole forest closed in.

These Saturday excursions to Ravensdale were the high point of the whole winter week, the point where dreams would converge and become real. Those dams and bridges in the streams were perhaps the nascent desire of an engineer, but I don't think so. They were rare constructive interludes in a dream world. Down in the valley was a proper river, and meadows, and the ruins of an old house – we called it a castle.

In spring the old house was surrounded by wild daffodils of a dark heavy yellow like buttercups, so strong a colour they were overpowering. Wild flowers were not too common. The taste of these daffodils and the early summer lightness of primroses ignited a passion.

> I wandered lonely as a cloud
> That floats on high over vales and hills
> … a host of golden daffodils.

Watching daffodils with poetry was a potent mix, a romantic moment to be at one alone with something wild. These flowers begat the love of poetry, read sparingly, a little at a time. This love has stayed with me since.

The ambivalence of these tensions meant that I had no childhood pals in Dundalk, just friends. I became a dreamer, not unhappy, but happily chaotic and retreating, when necessary, into a dream land which has protected me ever since.

The positive happy pole was Gyles Quay. Just after the war, and especially after my grandfather had died, we spent every summer at Gyles Quay. There my father could indulge his passion for sunbathing and long walks, and we the children roamed free. Sunbathing in Ireland is a haphazard pastime, you must be able to take advantage of every bit of sun. So we went for the summer holidays to Gyles Quay, not a fashionable seaside resort, but a stretch of flat, usable sand and a pier with coastguards' houses and some twenty houses scattered around, all private and secluded.

Gyles Quay was 10 miles from Dundalk, to the east, and could be reached by train or car. Here we were all summer people. There was no sense of defensive ownership. Nobody was rich, or excessively poor; we were equal. Families came back year after year and we were fully paid-up members of the clan. I loved Gyles Quay. The weather was often cool and wet. I preferred it that way. The sea was cold but clean and blue, and wonderful for swimming.

Swimming was freedom; through the waves out to sea or along the pier, slow, methodic expenditure of energy, an uncompetitive oneness with the rhythm of the sea. I was the best swimmer there, it was my forte and I was admired for an even elegant stroke, among people most of whom were terrified of the water. The others who swam were not very good; nor in truth was I, but I looked good and that was all.

Gyles Quay, on the Cooley Peninsula, was also near the border. But here that was a distant silent place we never talked about, and it was without menace. The occasional long warm summer days were almost perfect, not too hot for a sensitive Irish skin.

The place filled up on sunny days with noisy happy families, eating sandwiches on the stony foreshore behind the sand and paddling intermittently in the cold sea. These visiting strangers filled me with delight because they watched me as I swam from the shore to the pier and back, like the Titans then conquering the English Channel. It became for a time my ambition – and my fantasy as I swam – to walk out of the sea at Dover covered in grease and in goggles, with everyone looking on in awe (especially in Dundalk and Gyles Quay when they would see my photograph in the *Irish Independent*). I don't remember if I pretended I was the first man from Ireland or just Gyles Quay to achieve this, but I was very proud of myself all the same. As I ran back home after my mammoth swim, all of 400 yards, I would smile and mutter, 'just training' to the admiring visitors. Nothing was impossible after such a swim. A physical well-being that required no temerity took over. It was something I was good at. Gyles Quay was where I grew up.

In the late evening sometimes we would go progging apples, searching them out from gardens and hedgerows. One night in a remote place we saw an apple tree in an old disused cemetery. The tombstones in the half light were menacing, but we went in. To get the apples I had to jump. As I landed the grave beneath me collapsed and I was two foot deep in the ground. Not for long. I realized how quickly I could run.

Each summer was an advance on all the previous ones. As I grew older my interests did not seem to change. I had no hobbies that I can remember, no special interests. Wandering and day-dreaming were about all I did, interspersed with spans of sport. There was an awareness of the opposite sex, an awareness that I was too timid and introverted to explore. There was an awareness also of my own growing sexuality. This, confined as it was within the rigorous rules of Mother Church, was more of a burden than a pleasure and something with which I was continually having problems.

The Church and its influence were everywhere, the principal framework for thought and exploration. It was the fountain of idealism, of the belief in goodness and of all the spiritual values. One heard of atheists and agnostics, indeed it was vaguely clear even then that my father was an agnostic, but he never talked about it and I never asked. Atheists and agnostics were so unreal that we could hardly believe they were serious. Not believe in God! How could they? Had they not read Aquinas? Meditation and sainthood were the ultimate ideals. But they, and the dreams about them, were constantly usurped by the ugly demon of sex, a battle I was not to win until I gave in to it much later.

In my early days, between the ages of about ten and fourteen, I was an altar boy, serving Mass as an apprenticeship to becoming a priest. I loved the ritual with its subtle rhythms and silences. The unchanging resonance of the priest's voice speaking the Latin, then coming to the points where we could join in and intone alongside him the Creed or the Our Father, was formal and satisfying, a detached thinking and advancing in belief.

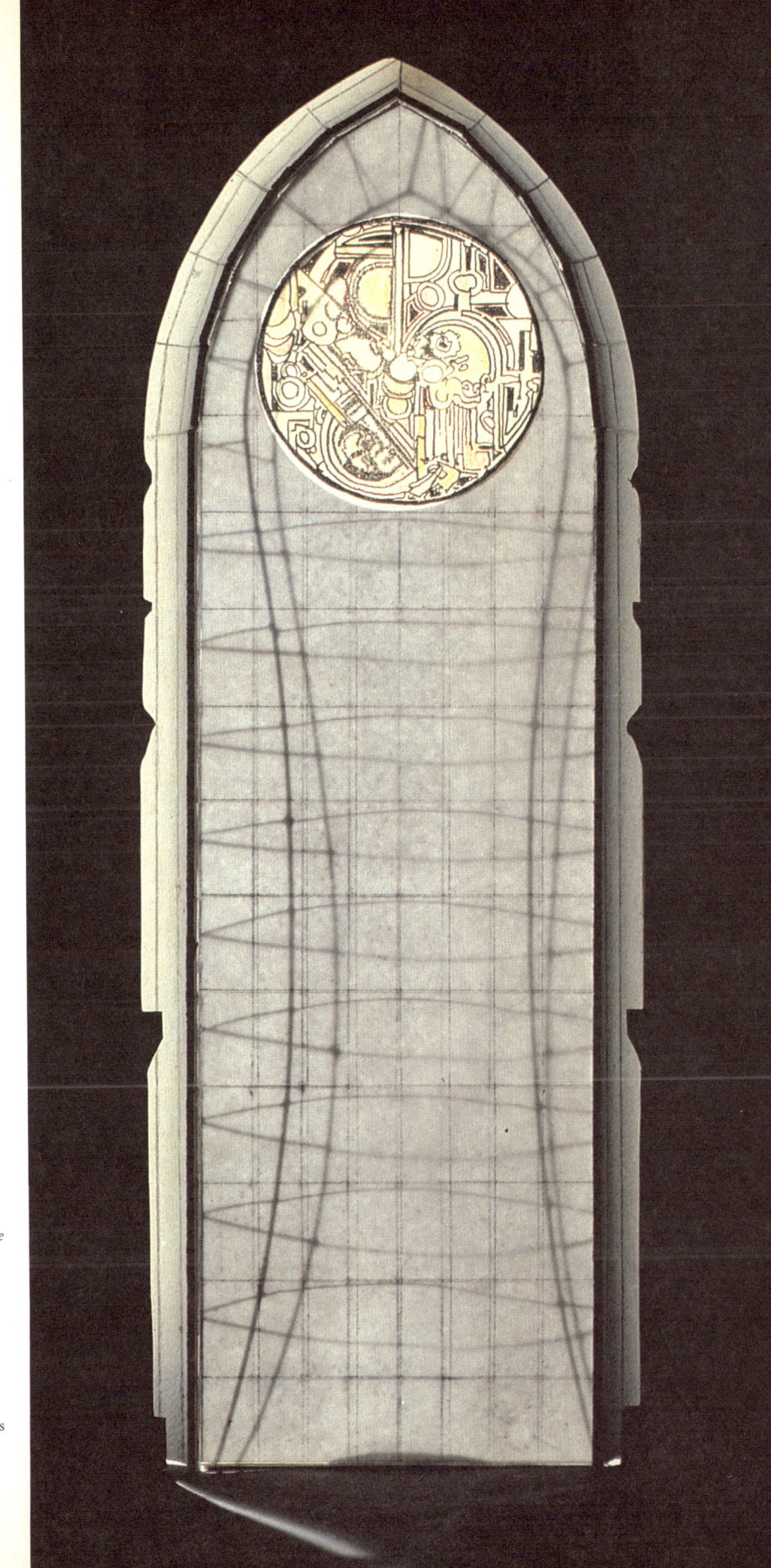

Translucent stone framed in stone Lille Cathedral, begun in the eighteenth century, was left unfinished. For the new west front, very thin marble or alabaster laminated on white glass would let light into a very dark space considered metaphorically as the Veil of Veronica. The central bay of translucent stone bonded to glass is suspended from a stone arch with a stainless steel tension bracing system, prestressed so the stone is always in compression.

For a time I became an altar boy in Masses for the dead, every morning at 8 o'clock in the small mortuary chapel, where the sweet smell of death and old women became almost addictive. I was fascinated by this close proximity to death, afraid yet drawn to participate in what was our ultimate destiny. It may sound morbid. It wasn't. It was more an inquisitive realization that death and life were the same, something you could feel and understand. In the Ireland of the 1940s, death was a cause for celebration. All the best parties were wakes and people often planned their own wakes in great detail, expecting to watch them from on high.

Later, at fifteen, I went to boarding school, Newbridge. My father, seeing my idealistic belief that I wanted to be a priest – oh, the joy my mother felt at this development! – wanted me to realize how wrong it would be. Instead of saying so, he sent me to be in close contact to see for myself. Newbridge was a school for farmers' sons, and I was almost the only boy there with pretensions to learning. Life was easy and I suppose boring. At Newbridge I had the benefit of a bad education. As I was bright it did not matter. It cured me of any desire to be a priest. It helped me find my level, not stuck up there in some power environment where life would have been a struggle.

It played one cruel twist of fate, though. By unfortunate coincidence the English teacher had a brother who had been sacked by my father. I did not know this. Sitting on the window, with the weir behind, he would mock my pretentious English style: 'Listen how the mathematician expresses himself.' Obviously if you were good at maths you could not write. That seemed sensible even to me. You could not expect to be good at everything. The possibility of writing was postponed for many long years.

After Newbridge I went to Queen's, Belfast. At that time, between the periods of IRA trouble, Belfast was a quiet, serious city. The choice was made by my father: 'You're good at maths. Engineering is the only job where you'll make money, and Queen's is the best.' So I studied engineering, starting with aeronautical engineering, but I could not stand the place where it was taught, so I changed to the civil engineering course. I didn't understand the engineering but I could do the sums. Understanding came later when I had started to work.

Jørn Utzon: sketch for
Sydney Opera House
Competition, 1957.

Sydney

After my primary degree and another year at Imperial College, I joined Ove Arup & Partners in 1956 to work on the Sydney Opera House. I worked with a small team under Ronald Jenkins, Ove Arup's senior partner. Jenkins was an engineer whose mathematical elegance and precision made a great impression on me. He represented a kind of ideal, an engineer who combined mathematical rigour with a clear structural understanding of how things worked. I worked for three years alongside him trying to find a logical solution to the conundrum of the Sydney shells. When Jenkins was replaced, Jack Zunz, who took over as the chief engineer, initiated work on a ribbed scheme solution in contrast to the previously researched structural shell solutions. The architect, Jørn Utzon, recognized this as the way forward. And I found myself one of the few people with an understanding of how to solve and codify the geometry of the shells.

As part of the agreement with Arups when I agreed to stay on the opera house project, I requested that I should go to Sydney to work on site; fortunately, I was allowed to go. This was after three years in London where, working with a large team, I did the principal analysis of the roof structure, both in its final built condition and during its construction; I also wrote the computer programmes for the geometry definition of the precast elements. When I got to Sydney I was to be the assistant to a very experienced engineer, Ian Mackenzie, who had been the supervising engineer for the podium structure. About one month after I arrived he fell ill and had to go to hospital. During his temporary absence I found myself in sole charge. And I discovered the work was really rather simple, no more than applied common sense. From then on and for the next three years, I was a resident engineer at the ripe old age of twenty-eight. In that position I became responsible, with a very talented surveyor, Mike Elphick, for the survey and positioning of all the shell and tile elements.

Memories of times past have a powerful influence. This is especially so when they concern people or things that have been important in one's development. Jørn Utzon is such a person for me. I did not work directly with Utzon; rather I was sprayed by the general influence of his philosophy and genius. I worked in his ambience for six years during the design development and construction of the roof of the Sydney Opera House. It was a long slow apprenticeship in the art of architecture, where there was sufficient time to observe

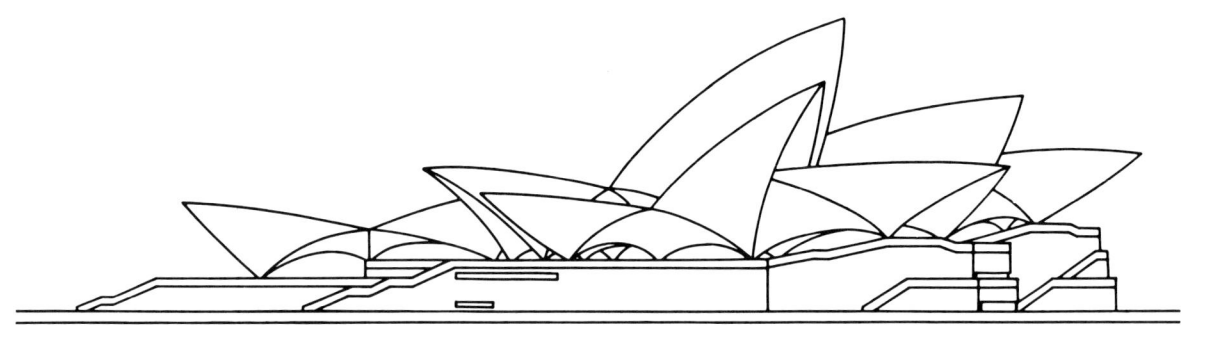

Sydney Opera House competition entry, east elevation.

Jørn Utzon with an early model of the Sydney Opera House.

Utzon's early main floor plan. His was the only entry which placed the two halls side by side on a single podium.

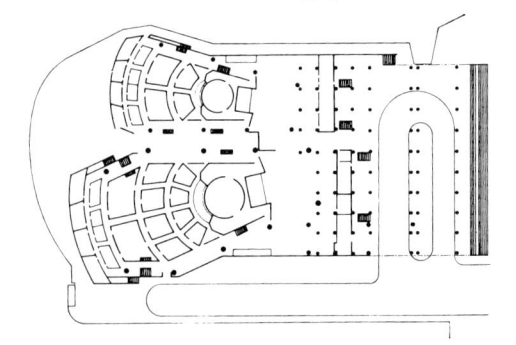

and to understand precisely the elements which contribute to making that building the masterpiece it is.

Before Sydney I had a very primitive appreciation of architecture. Life in rural Ireland in the 1950s had given few clues to what it was all about, so I came to the experience innocently, like blotting paper ready to absorb any information which came my way.

The design of the opera house was a dramatic story from the beginning. The Opera House Competition in 1957 was one of the first international architectural competitions after the war. Its integrity, something for which international competitions then were not renowned, was guaranteed by the quality of the architects on the international jury. This was dramatically demonstrated by the possibly apocryphal story told of the way in which Utzon's scheme was selected.

The story goes that when the international architect members of the jury arrived in Sydney they asked to see the entries which had been rejected by the technical panel. The architect Eero Saarinen asked that Utzon's scheme be reinstated, and then, in spite of the paucity of the information, it was unanimously chosen as the winner. Like Piano and Rogers' scheme for Beaubourg later, it had a number of features which separated it from all the other entries. It was the only one to place the two halls side by side on a single podium. It was also the only scheme which had the entrance from behind, entering around the stage, and with galleries and bars facing the magnificent harbour views. And then there were the shells. These were powerful sail-like structures sitting on the podium, visually balanced one against the other, one set covering the large hall, a smaller similar set covering the small hall. I had always imagined that the ridge outlines were inspired by the form of decreasing sound waves. The simplicity and magic of the scheme survived all the trauma and these two elements, the podium and the shells, have come together to form one of the dominant architectural symbols of our time.

Soon after winning the competition, Utzon started working with Ove Arup, another Dane (although he was born in Newcastle upon Tyne). Politics dictated that the building should start on site almost immediately, long before the scheme had been properly designed. Many of the subsequent problems have been blamed on this rushed start, but perhaps the whole thing would not have been possible otherwise, because from the beginning the team had the oxygen and

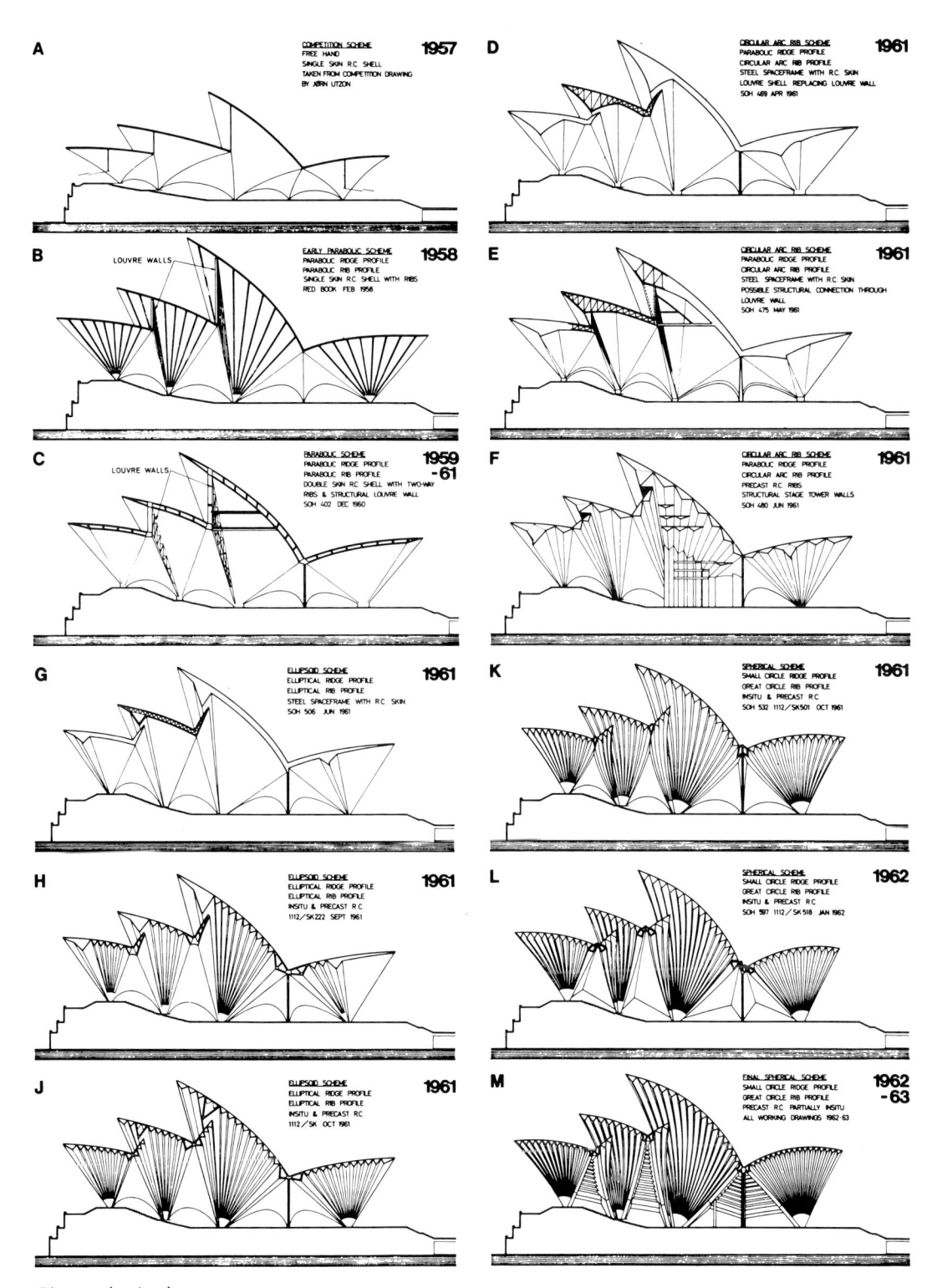

A	COMPETITION SCHEME — 1957 FREE HAND SINGLE SKIN R.C. SHELL TAKEN FROM COMPETITION DRAWING BY JØRN UTZON	**D**	CIRCULAR ARC RIB SCHEME — 1961 PARABOLIC RIDGE PROFILE CIRCULAR ARC RIB PROFILE STEEL SPACEFRAME WITH R.C. SKIN LOUVRE SHELL REPLACING LOUVRE WALL SOH 469 APR 1961
B	EARLY PARABOLIC SCHEME — 1958 PARABOLIC RIDGE PROFILE PARABOLIC RIB PROFILE SINGLE SKIN R.C. SHELL WITH RIBS RED BOOK FEB 1958	**E**	CIRCULAR ARC RIB SCHEME — 1961 PARABOLIC RIDGE PROFILE CIRCULAR ARC RIB PROFILE STEEL SPACEFRAME WITH R.C. SKIN POSSIBLE STRUCTURAL CONNECTION THROUGH LOUVRE WALL SOH 475 MAY 1961
C	PARABOLIC SCHEME — 1959-61 PARABOLIC RIDGE PROFILE PARABOLIC RIB PROFILE DOUBLE SKIN R.C. SHELL WITH TWO-WAY RIBS & STRUCTURAL LOUVRE WALL SOH 402 DEC 1960	**F**	CIRCULAR ARC RIB SCHEME — 1961 PARABOLIC RIDGE PROFILE CIRCULAR ARC RIB PROFILE PRECAST R.C. RIBS STRUCTURAL STAGE TOWER WALLS SOH 480 JUN 1961
G	ELLIPSOID SCHEME — 1961 ELLIPTICAL RIDGE PROFILE ELLIPTICAL RIB PROFILE STEEL SPACEFRAME WITH R.C. SKIN SOH 506 JUN 1961	**K**	SPHERICAL SCHEME — 1961 SMALL CIRCLE RIDGE PROFILE GREAT CIRCLE RIB PROFILE INSITU & PRECAST R.C. SOH 532 1112/SK501 OCT 1961
H	ELLIPSOID SCHEME — 1961 ELLIPTICAL RIDGE PROFILE ELLIPTICAL RIB PROFILE INSITU & PRECAST R.C. 1112/SK222 SEPT 1961	**L**	SPHERICAL SCHEME — 1962 SMALL CIRCLE RIDGE PROFILE GREAT CIRCLE RIB PROFILE INSITU & PRECAST R.C. SOH 597 1112/SK 518 JAN 1962
J	ELLIPSOID SCHEME — 1961 ELLIPTICAL RIDGE PROFILE ELLIPTICAL RIB PROFILE INSITU & PRECAST R.C. 1112/SK OCT 1961	**M**	FINAL SPHERICAL SCHEME — 1962-63 SMALL CIRCLE RIDGE PROFILE GREAT CIRCLE RIB PROFILE PRECAST R.C. PARTIALLY INSITU ALL WORKING DRAWINGS 1962-63

Diagrams showing the development of the geometry and structure of the shells for the Sydney Opera House.

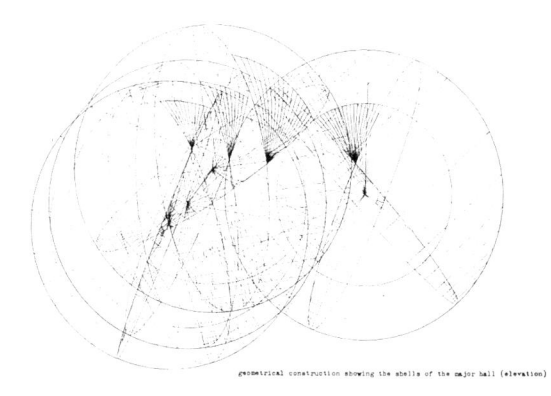

The breakthrough came when Utzon changed the silhouette so that the ribs of the shells could all be taken from the same sphere.

geometrical construction showing the shells of the major hall (elevation)

Opera House construction site on Bennelong Point showing casting yard for rib and arch segments.

Rib segment being cast.

incentive of a building under construction to drive them on. With such an innovative project the doubts arose often but there was no possibility of stopping.

The gradual development of the solution for the shells illustrates very well the inventive thinking of the engineer working with the architect. Utzon had a great ability to get the best from those around him. In the beginning the engineers attempted to build exactly what he had drawn in his initial competition sketches. The form was dominated by the ridge profile. But this did not conform to any regular geometry. A geometry was found but it was established parametrically, rather than being from a regular geometry. Organizing the tiles on the outside surface presented a major problem. Quality was the by-word. But quality control in an *ad hoc* on-site tiling sequence was not really possible on the scale of these buildings. Quality control of the tiling required precast tile elements which in turn required a regular geometry. A rethink was needed.

In my opinion the rethink began with the realization that the shells were three-dimensional objects, seen from close up and far away, not the quasi two-dimensional forms dominated in elevation by the poetic ridge outlines. The change when it came was dramatic. Everything was made from parts of the same sphere. The outer surface was divided up into a set of regular, chevron-shaped tile lids, enabling simple moulds to cast repetitively the whole of the outer shape. The structure too was precast in identical circular ribs, cut on the great circles of the sphere and chopped off by the ridge plane. The articulation of the inner surface enabled the glass walls and all the other items to be attached to the inside in a natural logical way.

Attention was focused on quality – quality of construction and quality of architectural detailing. The tiling illustrates this very well. There are two kinds of tile: a glazed reflective self-cleaning white tile, and a matt cream-coloured tile. The outlines of the main shells are defined by great bands of cream tile lids so that the articulation of the surface can be seen from afar, from the harbour on your first sight as you enter the main section. Then as you get nearer, each tile lid, on average 2.5 metres by 1.5 metres in size, is outlined by a single band of cream tiles. And then as you stand next to the building, each tile is articulated by a recessed joint, which catches the sun and defines the curvature of the surface.

At an even more detailed level the tiles, which were extruded and cut to size, are orientated so that they all lie the same way, with the

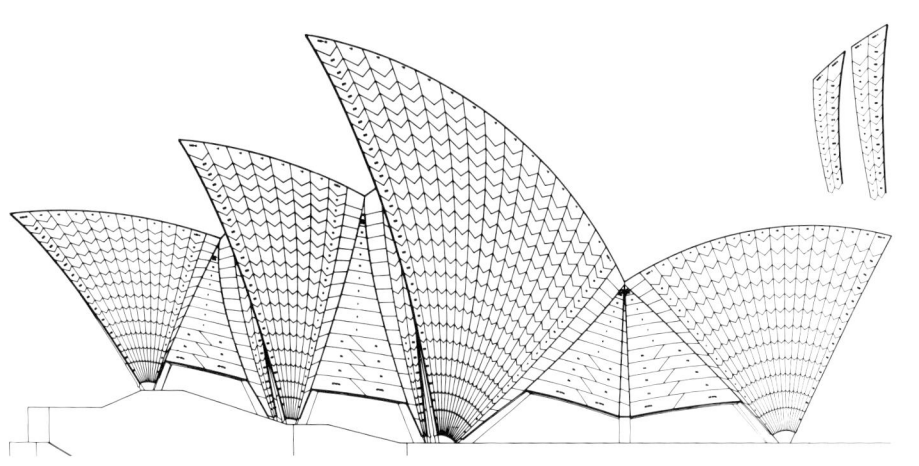

Sydney

Drawing showing organization of tile lids on shell surfaces.

cut edge picking up the tangential sunlight and adding sparkle to the surface.

That the scale of the building works at every level, providing interest and articulation from wherever you are, is a key lesson I was to remember from my days in Sydney. It was this influence – the awareness of the importance of the integrity of the building's construction and of the need to provide interest and articulation in the structure at all levels – which led directly to the proposal to use cast steel in the Centre Pompidou, and hence to the gerberettes. By using this technique the building does not dominate. It is the details which control the reaction of the public and hence their perception of the scale and warmth of the building. It is a return to the interest and romanticism of Gothic architecture, with its great scale where the *trace de la main* is still visible.

The influence of such an important period cannot be encapsulated in a single feature of the opera house. Walking around during construction, the variety and complexity of views of the half-completed building were a constant wonder. Many of them were the accidental product of the building's solution, but to a young naive engineer they were a magical endorsement of the time, effort and pain that the construction brought to many. They confirmed that if something was worth doing, it was worth doing well.

Of course memory is selective. I remember how occasionally I would find Utzon on site explaining some aspect of the building to the minister (the client) or some other visiting dignitary. Like the day I came across him telling the minister why the back wall of the stage that they were standing in front of was not really there. It was big, 20 metres by 30 metres blank, and confronted you as you emerged into the entrance under the great south-facing shell. The minister, being reasonable and prudent, had suggested that it was about time to commission an artist to do a mural or a statue. But to Utzon it was quite clear. The wall was not there. The flow around the sides was what counted. The wall had no presence. Listening, I was convinced. I understood how something so big and dominant could be said not to exist. And the minister must have been convinced too because no artist was ever commissioned.

The resignation of Utzon from the opera house project was of course political. An untenable situation was created in which Utzon inevitably resigned in order to demonstrate, he thought, that the architect was indispensable to the project. But the politicians moved

Typical tile lid, prefabricated on Bennelong Point.

Tile lids being cured and stored in casting yard.

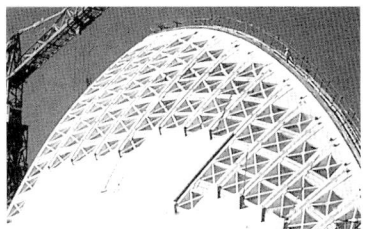

Roof shells being clad with tile lids. Bolts stick out of rib segments for the later attachment of concrete lids carrying tiles.

An Engineer Imagines

Roof under construction. The hollow concrete roof ribs which form the shells are exposed before being clad with tiles.

quickly and he was defeated. Naivety perhaps, but an architect can hardly be blamed for that and Utzon was not the first to make that mistake. It shows how all-encompassing the talent of an architect must be.

Maybe there was another reason. By then Utzon had been living with the project for almost nine years. He had come to Sydney and, cut off from his usual inspiration and background, he had struggled to refine his design. There was no effective client. The opera house had been the dream of one man, Cahill, then the Premier of New South Wales – a symbol for Sydney. But an opera house requires an opera company to help establish the brief and to act as arbiter in matters of design. The opera, ballet and theatre companies involved lacked the power to act as users of the building. And the building was way over the original cost. The money was there – the Australian love of gambling and the Opera House Lottery ensured that – so the cost became a political issue. I think Utzon was tired. He needed to recharge himself and it was at that moment that the politicians pounced.

And then there was his charm. Utzon was tall, elegant and impossible to resist. He could draw and sketch like a genius. He could have his way on anything. Charm, as Evelyn Waugh has noted, is 'the disease of a cold damp climate'. Ireland too has charm. I've watched it there. It corrodes what it touches. It saps energy because everything is so easy. And when it came to the crunch Utzon had estranged the friends he needed. Perhaps he had charmed them once too often.

It is interesting to speculate on what might have been. Within a context of increasing conflict between the architect and the government client, the final break was sparked by a dispute over the acoustic ceiling and glass walls at the ends of the building. Utzon wanted to develop a new material, plywood reinforced with aluminium. The government wanted a competitive tender. The integrity of the structure appeared to be at stake. Utzon could not help defending that, because an inner architectural integrity prevailed in everything he did, particularly with materials.

Utzon lost that battle because he could not compromise. In many walks of life that is a major fault. But for an artist, if an architect can be called an artist, it is perhaps the finest epithet of all.

After Sydney I spent eighteen months in America, six months in New York and twelve months at Cornell where I was a visiting

Interior view of a shell during construction.

Exterior view showing intersection of shells. Each tile lid carries cream-coloured matt tiles along its long edges with white reflecting glazed tiles over its central area.

An Engineer Imagines

The Opera House after completion.

scholar, a post which enabled me to sample without pressure the wide range of the university's courses. This sojourn enabled me to reform my ideas on engineering and prepared me for the years ahead. I returned to England in 1968, and I worked with Structures 3 at Arups. The principal work was with Frei Otto, during which time I first started to understand non-linear behaviour. And hence to Beaubourg.

Further reading on the Sydney Opera House:
Arup Journal, October 1973, Sydney Opera House Special Issue
'Sydney Opera House', paper published in *The Structural Engineer*, vol. 47, no. 3, March 1969, pp. 99–132.
Discussion on the 'Sydney Opera House', paper by O N Arup and G J Zunz, published in *The Structural Engineer*, vol. 47, no. 10, October 1969, pp. 419-425.
A detailed explanation of the events leading up to Utzon's resignation is to be found in John Yeoman's *The Other Taj Mahal* published by Longman, Green & Co. Ltd (London and Harlow, 1968). The specific dispute over the acoustic ceiling is to be found on pp. 137–147.

View of Sydney Opera House, Harbour Bridge and North Shore.

4

Ove Arup the man

I could not give a true picture of the influences which have formed my present attitudes without mentioning Ove Arup, my father in engineering.

When I joined Arups in 1956 Ove Arup was already 'the Old Man', a vague and venerable figure who floated above and around a young aggressive, ambitious organization. For a young engineer, lost in the vastness of London and cocooned in layers of seniority, the name evoked a sense of myth. I saw him occasionally, at the Christmas party, at the summer outing, a tall patrician figure, detached and kindly, watching benignly as the young ones enjoyed themselves. You could feel a sense of fun, an impishness that belied his position. It was the time of Hemingway's *The Old Man and the Sea*, a book which had impressed on me the remoteness and detachment of age, the distillation of wisdom and purpose which old people embody. It all seemed to fit.

I had joined Ove Arup & Partners because I had heard that it was a place where an oddball could fit in. Engineering then was a very serious profession. Perhaps it still is. Engineers were expected to know what they were about, to have a natural feel for their profession. I was an engineer by accident, tentatively feeling my way to a career, without any natural instinct for engineering. The atmosphere of Arups helped me survive.

Where then did this atmosphere come from? Clearly 'the Old Man' was the fountain.

Ove Arup defined an attitude, an integrity towards what one did which permeated down from his remote and distant sixth-floor office. Occasionally one would get called to work with him on some project. But that was difficult, and by the time I joined Arups no real contact was possible. That was reserved for those who had worked with him when he was more active. He presented a benign philosophical exterior and wanted principally to talk of the great issues of the day. His concerns were catholic. He felt that engineers did not take their responsibility with enough awareness of their role in society.

Towards the end of his life he wrote a short manifesto of his concerns, which I have taken as my own, as though his concerns were handed on for us, the following generation, to act on. Its central feature was the human predicament at the end of the twentieth century. It is best articulated in two extracts from lectures given at the end of his life.

An Engineer Imagines

> In the past the environment, the landscape in all its natural and urban forms just happened, it was never before deliberately created by man, except in small patches. The technological revolution is changing all that. Man's battle with nature has been won. Whether we like it or not, we are now burdened with the administration of the conquered territory. Nature reserves, landscape, townscape: they will all be wantonly destroyed, to the ultimate ruin of man, or they must be deliberately planned to serve his needs. Much has been destroyed already and more will be destroyed, but the alarm has sounded. Pollution, the population explosion, these things are news. The battle is on and it is a crucial battle for mankind. Those who long to return to the good old days must be told that that road is now closed.[1]

The lecture embodies Ove Arup's developing philosophy, as succinctly relevant today, as when it was written. The lecture has that self-deprecating ironic wit so typical of Arup the person. Parts of it are pure Ove.

> The Modern Movement ... discovered that the work of bygone engineers was in fact architecture. It is now accepted that bridges and factories and all that are architecture. So is housing, in fact everything built is architecture. And the same spirit which is supposed to be moving architects is behind town-planning and landscaping as well as interior design and furnishing. Everything made by man for man's use now has to be designed. And in all these spheres dedicated engineers are trying to conjure forth that mystical spiritual quality which is the essence of art.[2]

As Arup grew older he became more and more concerned with the predicament of humanity at the end of the twentieth century. Perhaps he realized that the lesson of his early years had been learned, of the need for architects and engineers to communicate, and that of course it had not solved the problem. An even greater challenge lay, and still lies ahead, the challenge of the environment, a challenge whose magnitude had only just become visible. The definition of this challenge preoccupied his last years: where and by whom must it be met? As engineers, are we but innocent followers of others or are we truly responsible? This was the theme of a statement made to the Fellowship of Engineering in September 1983, when Ove Arup was eighty-eight years old:

1,2 Extract from 'Architects, Engineers and Builders', the Alfred Bossom lecture of the Royal Society of Arts, 11 March 1970.

> ... it is my conviction that whilst we have become very clever at doing almost anything we like, we are very backward in choosing the right things to do. This is, of course, taking a global view of the behaviour of mankind and that, I submit, we are simply forced to do in view of the tremendous power for good and evil conferred on us by our sophisticated technology. It has brought us tremendous blessings, and it has also done tremendous damage to our planet and its inhabitants ... And as mentioned, the decision about how to use it is not generally made by the engineers. But engineers are world citizens as we all are and as they are largely represented on the design teams preparing the designs, which determine what is made, they are in a good position to judge the consequences for mankind of proceeding with doing what we are about to do. Would it not be a good thing if they had a say in what we should do and have they not a duty as citizens of the world to warn us of any dangerous consequences which would result from our action?[1]

And he ended with a plea. It is remarkable, but at eighty-eight all the passion of his youth was still there:

> My only hope is that this well-educated minority will swell to include the less well-educated majority so that even governments can start to think about how to alter course without creating world-wide chaos. It will be extremely difficult. It must be a slow and controlled process and its success depends on whether we can convince a majority of our leaders and their followers that we need to alter course. Doing a 'U-turn' in the mid-stream of traffic is dangerous, we can hardly avoid severe trouble and hardship. We are not helped by fanatic peace-mongers, feeding on simplistic slogans, who think they can achieve universal peace by hate and destruction. Pulling down is easy, building up is difficult. We have to employ slogans which the great mass of people can understand and support, but they should appeal to their good instincts, not their bad ones. This is a source which is not so often tapped by our politicians, but I believe its power could be overwhelming if our leaders had the courage to build on it. Ideals must be tempered by realism but should not be poisoned by cynicism or hate. In the end all depends on our own integrity.[2]

Ove Arup did not rest on his laurels. He did not stop and survey the kingdom he had created. He worried on our behalf. He was true

1,2 Extract from statement to the Fellowship of Engineering, by Ove Arup, September 1983, unpublished.

to himself to the end, a remarkable achievement for such a long and varied life.

If, in writing this book or in anything I have done, I have helped to solve the problems Ove so eloquently identified, I will feel that it has been an achievement to be proud of. He guides me still.

Ove Arup, 1895–1988.

5

The Role of the Engineer

I am an engineer. Often people will call me an 'architect engineer' as a compliment. It is meant to signify a quality of engineer who is more imaginative and design-orientated than a normal engineer. This is because in the minds of the public and of other professionals, the engineer is associated with unimaginative dull solutions. If people find an engineer making original designs, designs which only an engineer can make, they feel the need to grant him or her a higher accolade, hence 'architect engineer'. It is not that I object to being called an architect engineer. Occasionally it may even be appropriate, but mostly it is not because there is a fundamental difference between the work and way of working of an engineer and that of the architect or designer.

To call an engineer an 'architect engineer' because he comes up with unusual or original solutions is essentially to misunderstand the role of the engineer in society. It is easiest to explain the difference between the engineer and others by comparing how each works and what they do. Designers such as the famous car stylists, like Pininfarina, or Giugiaro, work essentially from within themselves. They respond to a design challenge by seeking to understand how they respond to the context and the essential elements of the problem: their response is essentially subjective. Different architects will respond very differently to the same problem. Their solutions will reflect their style preference and their general belief in an appropriate response to a problem. Thus, if you ask a particular architect to respond to a design challenge, he will always give a solution based on the classical order, a solution which reflects his belief that the classical order is the only satisfactory response which preserves an urban sense of scale and recognizes a link with the past which he interprets in this way. Other architects would choose a different approach based on their subjective reactions as to what an appropriate architectural response should be. Both the likely responses would be known beforehand and would probably have been an important factor in the selection of the architect in the first place. An architect's or designer's response to any design challenge is subjective and is based on his feeling on the correct and appropriate response. He is employed to express his personal view of the correct solution.

An engineer and an architect would rarely find themselves tackling the same kind of problem, but it does happen, as I will demonstrate later in this story. The engineer when faced with a design

1960 Ferrari Coupe 250/GT designed by Pininfarina.

Lloyd's of London, detail of concrete bracket.

Lloyd's of London, view of concrete structure in atrium.

View of a Grande Serre at the Musée des Sciences et de l'Industrie at Parc La Villette, Paris.

challenge will transform it into one which can be tackled objectively. As an example, the engineer might seek to change the problem into the exploration of how to exploit a particular material completely within the context of the architecture.

Thus the Lloyd's of London building became an exploration of the use and properties of concrete. And the engineer's contribution was to try and make the structure an essay on expression in the use of concrete. But it was the properties of concrete which motivated the search and the solution.

Similarly at the La Villette 'greenhouses' in Paris. The architect defined the architectural intention, and the engineer transformed the simple architectural statement into an essay on the nature of transparency and of how to use the physical properties of glass to convey fully the concept of *transparence*. As an engineer I worked essentially with the glass. It was the properties of the material which motivated the development of the design. Thus, although we can say that there was originality and aesthetic choice in the way that the design developed, this way forward was directed by the need to express the properties of the glass in full.

I would distinguish the difference between the engineer and the architect by saying the architect's response is primarily creative, whereas the engineer's is essentially inventive.

The architect, like the artist, is motivated by personal considerations whereas the engineer is essentially seeking to transform the problem into one where the essential properties of structure, material or some other impersonal element are being expressed. This distinction between creation and invention is the key to understanding the difference between the engineer and the architect, and how they can both work on the same project but contribute in different ways. Indeed, now it is important that engineers start to educate both people within the profession and the public at large on the essential contribution that the engineer makes to even the most mundane project. To begin that process, and to understand fully the problems we engineers face, I would now like to examine the situation of the engineer in general.

Engineers to many people, especially to the public, are mysterious figures. The most frequent remark is: 'What do they do? They just makes things stand up,' as though this were not a noble thing to do. No, in our media-dominated society it is the image not the content which matters. But the engineer's role is crucial nonethe-

Kurfürstendamm 700, Berlin
The design of the smooth, gently curving and tilting skin is based on an adaptation of a conventional structural curtain wall, where the grid is a structural mesh of aluminium extrusions suspended from the top floor in a continuous sheet. The sliding connections to lower floors must restrain wind loads lateral to the glass but not vertical loads.

The Role of the
Engineer

less, even in matters of image. The problem is that, in the simple world that the media favours, the role of image-making is given to others – to designers, for industrial artefacts such as cars, household goods, and so on; and to architects for the monuments of our built environment. It is not that there is anything wrong with this approach *per se*, it just ignores the vital role played by engineers in the creation of all the things that are built or made today.

'Engineer' is not a proud word in our language. It is not a word to make you stand up and beat your chest. Sometimes it is associated with electricians or other craftsmen from unionized labour. This distortion at the bottom end of the scale is the prevalent thought in the mind of many in Great Britain, although this is less so in France, Italy, Spain, or Germany. The engineer is not identified with other professions, such as lawyer, doctor, architect. The word 'engineer' offers no protection. Anyone can say he is an engineer – not a chartered engineer, but the distinction is fine, and hardly one that the general public could be expected to understand. And even those who do understand have difficulty in understanding what the engineer actually does. How then does this come about?

Partly it may be explained historically. But the provenance of the engineer in the nineteenth century is noble indeed. Telford, Stephenson, Brunel father and son, Eiffel, these are names of pride and achievement, names that any profession would welcome as their antecedents. Indeed in their time they outshone architects and other designers and their achievements are still spoken of with awe.

Where have we gone wrong? Are we somehow different today, doing less exciting, less urgent, less vital tasks in our complex modern world? But then we have sent men to the moon. Grey people, you say, working in teams, not great individuals placing their personal authority on the line in what they design. But is that true? Of course not. Every team has a leader, one who is ultimately responsible for the work that the team does. Remember the O-rings on the Challenger space shuttle, or the engineer responsible for the DC10 which crashed near Orly in 1974. There it was clear what the engineer's role was. The problem, then and today, is that the name and role of the engineer are not known to the public. Engineers work incognito. Unlike their predecessors, today's engineers work behind a screen of other egos. The great nineteenth-century engineers were of course entrepreneurs as well as engineers, financing and being financially responsible for the artefacts they constructed.

The Severn Bridge.

The Conway Crossing, construction sequence. Concrete tunnel tubes are cast, floated to site position and sunk. The water is then evacuated.

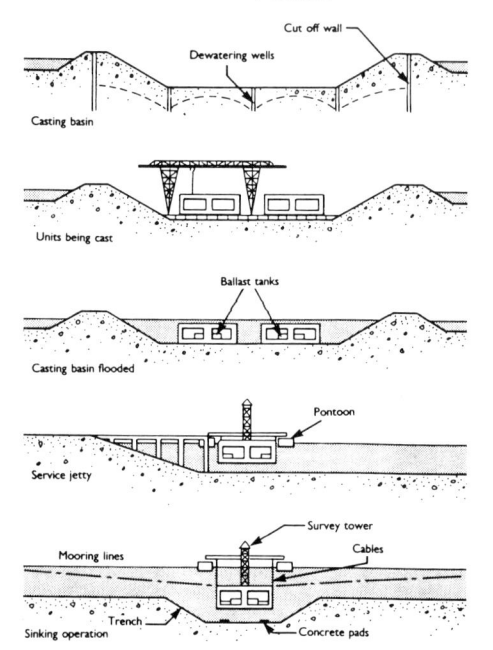

Engineers then, as engineers now, were doing the same job and taking the same responsibility.

The engineer's role in the design of large structures is easy to understand. Take the bridge which carries the M4 motorway across the River Severn. That bridge in its time was highly innovative. It was much lighter than other bridges being built at that time, such as the Verrazano Narrows Bridge in the USA, and it spawned a completely new way of making very long suspension bridges. The Humber and Bosphoros bridges are two bridges made by the methods pioneered in the Severn Bridge. The invention and innovation shown in these designs came from the engineers Freeman Fox of London. Another example of recent inventive engineering, not so visible but as original, is the Conway crossing in north Wales. Here the engineers made two concrete tubes, in precast elements, joined together on the surface, which were sunk into two trenches pre-dug on the estuary floor. By evacuating the water from the preformed tubes an underwater road tunnel was created.

I include these two recent examples of the work of engineers to show that today's engineers are just as daring, just as inventive as their Victorian counterparts. It's just that even when we know about their work, we don't have a label to attach to the work.

I believe that it is this lack of a personality to identify with the work which is the fundamental weakness of the engineer's position. Engineers need identity. Engineers need to be known as individuals responsible for the artefacts they have designed.

But what is also needed is to communicate some of the excitement of engineering. The truism of engineering is to say that engineers have, as their challenge, the conquering of the great and varied forces of nature. And that is of course true. Gravity, wind, snow, earthquake are all parts of the challenge. The essence of the engineer's role as a designer is to conceive works to withstand whatever forces the natural environment and people place upon them. Other engineers – water engineers, environmental engineers, foundation engineers – all have great primeval forces to withstand in their designs. Assessing these forces and being sure that they have been properly accounted for is far from automatic. Codes and tradition give a guide but they are just the beginning. They do not cover specific situations, and therefore the engineer is always left to make the final assessment.

Engineering is challenging and exciting and requires the highest skill. And it is glamorous. It is just that we fail as engineers to explain what we do and how we do it. It is important in all this that we recognize and explain that there are many decisions that only an engineer can make.

The gerberette at Beaubourg is an example of an apparently architectural decision that could only have been made by an engineer. It is this special role of the engineer which needs to be better understood. We have first to understand how this role differs from that of the architect. Broadly, the architect responds to the site, the context of the situation, and creates an emotional response to the situation. It is his personal assessment.

The essence of the engineers' problem then is that their work is either not understood or is given scant treatment by the media; even the engineers' own media fail fully to express the mystical excitement of the engineering challenge.

However, within the context of the work which engineers do themselves, there are certain features which make the role of the engineer sharper. An engineer must not be wrong, because human life and human safety are dependent on the engineer's work being right. That is the bottom line.

In this general world the engineer's work is not understood and is not valued for the contribution it makes to even the most mundane of artefacts. I want to show, principally through the work I have done as an engineer, the scope for inventiveness and innovation that exists, and also to identify how the engineer's contribution can enhance the architecture of many artefacts.

Of course I do not wish to imply that all engineers are somehow unappreciated geniuses waiting for the opportunity to express their inventive skills. This is clearly not the case. The argument is that there are many engineering contributions which go unrecognized, or which are attributed to the architects or others with whom the engineer is working.

Many engineers are themselves affected by the general expectation that society places upon them and they behave accordingly. Indeed they may in their pragmatic way encourage and foster the very atmosphere that inhibits them when they wish to be inventive. This I call the 'Iago mentality'.

There is an essay entitled 'The Joker in the Pack' in W H Auden's collection *The Dyer's Hand*. In it he examines the role of Iago and his unfailing rational arguments in the destruction of romanticism.

Iago (David Suchet) reasons with Othello (Ben Kingsley), Royal Shakespeare Company production (1985).

In *Othello*, Iago uses sound, sensible arguments throughout to destroy the romantic idyll of Othello's existence.

Iago, as the agent of rational argument, undermines those fragile characteristics of love and loyalty by the constant application of simple rational argument. Science, Auden argues, also destroys our romantic and artistic creativity by constantly requiring us to pass the test of rational acceptance. In the dialogue of architecture and engineering, the engineer is the voice of rationality and reason. It is a role which is all too easy to play. After all, one is being sensible, reasonable, modern in questioning the more far-fetched flights of fancy of some architectural proposition. This then is the engineer's principal destructive weakness, to play Iago to the architect, or indeed to another engineer, Othello. In Alexander Pope's phrase, what the engineer can do is to 'Damn with faint praise …' and in so doing destroy the fragile shoots of creativity in others. It is of course not only engineers who can be part of the destructive process. Richard Weinstein, who first introduced me to Auden's essays, held as his thesis that much of the modern architectural development which relies upon scientific correctness to justify its decisions is in the grip of the Iago approach. Such a position obscures the individual contribution of the architect and buries it in apparently rational justification. And although this is not the work of the engineers, it uses adherence to engineering principles to justify its choices.

Is there any escape for the engineer in this world of rational absolutes? I believe that there is. Let us examine again the role that engineers play, or at least can play. They work with materials. They can work with light. They can work with air. They work with the basic and fundamental elements of construction. They work with the content, not the image. It is a truism that most people find modern architecture cold and alien. Whereas a Gothic cathedral will express the real and physical presence of the stone from which it was made, and of the masons who laboured over its construction so many years ago, very few modern buildings carry the same physical presence of the materials of which they were built. In short, the *traces de la main*, the evidence of those who built it, is not there. These buildings are not tactile.

This is the positive role for the engineers' genius and skill: to use their understanding of materials and structure to make real the presence of the materials in use in the building, so that people warm to them, want to touch them, feel a sense of the material itself and of

Translucent stone framed in stone Lille Cathedral, begun in the eighteenth century, was left unfinished. For the new west front, very thin marble or alabaster laminated on white glass would let light into a very dark space considered metaphorically as the Veil of Veronica. The central bay of translucent stone bonded to glass is suspended from a stone arch with a stainless steel tension bracing system, prestressed so the stone is always in compression.

Kurfürstendamm 700, Berlin
The design of the smooth, gently curving and tilting skin is based on an adaptation of a conventional structural curtain wall, where the grid is a structural mesh of aluminium extrusions suspended from the top floor in a continuous sheet. The sliding connections to lower floors must restrain wind loads lateral to the glass but not vertical loads.

The Role of the Engineer

the people who made and designed it. To do this we have to avoid the worst excesses of the industrial hegemony. To maintain the feeling that it was the designer, and not industry and its available options, that decided, is one essential ingredient of seeking a tactile, *traces de la main* solution.

A building does not have to be made of brick or stone to achieve this, but rather it is the honesty and immediacy in the use of its principal materials which determines its tactile quality. That was the essential reason for the use of cast steel in Beaubourg, and for many other choices which will be explained elsewhere.

This then is a noble role that the engineer can assume – the role of controlling and taming industry. The building industry has an enormous investment in the *status quo* and, like Iago, will use every argument to demonstrate that other choices are irrational and not very sensible. Only the engineer can withstand these arguments, demonstrate the wrongness of the position of industry and demolish its arguments. In this scenario, the engineer becomes critical and can save his soul.

In general, though, the most powerful way that an engineer can contribute to the work of architects is by exploring the nature of the materials and using that knowledge to produce a special quality in the way materials are used.

Exploration and innovation are the keys. I have noticed over the years that the most effective use of materials is often achieved when they are being explored and used for the first time. The designer does not feel inhibited by precedent. For example, there is little to compare with the expression of the nature and character of concrete form in the bridges of Maillart. He made forms and shapes which fully reflect the way concrete is made and the plastic forms that can be achieved with it. Maillart's concrete compares with Brunel's use of steel or Telford's use of iron.

In any of these structures there is a simple honesty which goes straight to the heart of the physical characteristics of the material and expresses them in an uninhibited way. This implies – and I believe it to be true – that when we can re-create as engineers that sense of adventure and innovation, we will be most successful in giving the tactile quality to our designs.

A design does not just have to be something you can touch; it can be tactile when a material is used to express its inner nature with feeling and is clearly the work of a designer who, in thinking about

1930, Salginatobel Bridge, Switzerland by Robert Maillard (1872–1940).

1864, Clifton Suspension Bridge by Isambard Kingdom Brunel (1806–1859).

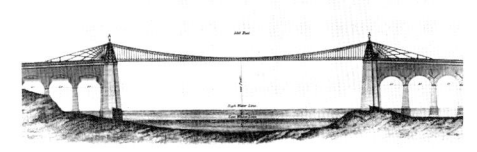

1826, bridge over the Menai Straits by Thomas Telford (1757–1834).

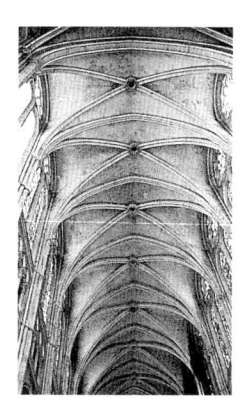

Vaulted ceiling of Chartres Cathedral.

the material, has made the perception of the material more real. Thus I believe that the glass walls at La Villette are tactile even though it is not possible physically to touch them.

Throughout this book I use the word 'tactile'. What do I mean by talking of architecture being tactile? Tactile quality is like empathy. It is like the feeling you get when you visit a particularly hallowed place. A friend recently described to me the powerful presence she felt when visiting Jerusalem, and especially when visiting the great religious places of the city. Generations of people have made their presence felt and left visible and invisible evidence of their presence. Gothic cathedrals, and many Renaissance palaces and churches, have this quality. This is helped of course by the fact that these were built of natural materials. But it is something more than that, some presence which puts you in contact with the building and the past. Animals too have this need to reassure themselves of others who passed before them when they go exploring or checking out a new place. They seek the scent of previous occupants. For that reason I believe that the primary quality which makes the built environment tactile is evidence that people have participated in its construction. It is for that reason, when I think of the alienation which exists today between people and the built environment, that I put it down primarily to the all-pervasive and essentially sterile role of industry, which suggests that decisions about this environment were made not by people but by the needs of industry.

The search for the authentic character of a material is at the heart of any approach to engineering design. This statement may seem excessive and even frivolous to many engineers. And it is important to emphasize that one should not invent and innovate just for its own sake. Innovation should have a real purpose and be contributing to the project. Nevertheless, some extra design element should be every engineer's objective whatever the project he is working on.

What about the engineer's basic obligation to build as cheaply and as economically as possible? I do not believe that economy and innovation are necessarily incompatible. Any project has cost constraints which must be met. They are a fundamental part of the design challenge, and finding a way to add one's special extra quality while respecting all the other parameters is one of the things which can make the challenge of design interesting and exciting to the engineer.

Stone gargoyles of Notre Dame Cathedral, Paris.

The Role of the Engineer

A double boom beam, Pompidou Centre, Paris.

Finally I would like to discuss the relationship of the engineer and the design concept. One day I was discussing with the architect Richard Rogers the fact that many of my contributions to design seem to arise as a complete concept. By this I mean the solution to the problem will often come fully worked out. I give two examples. The main beam in Beaubourg, with its double top chord and double bottom chord and the cast nodes which connect the double booms to the single-layer sheer members, appeared fully formed in my mind one night in bed. I guess that concentration on the search for a solution over the previous weeks had defined the problem very clearly. Nevertheless this complete beam form was there when I awoke, and it did not change in any important aspect during the subsequent architectural development. It even had what became one of the basic design rules for the Beaubourg steelwork built into the solution. This was the use of solid rounds for tension, tubes in compression members and castings for the node. How did this happen? I am not too sure. What I do know is that the decision to use double booms came from the belief that some light passing between the booms would lighten their visual impact very considerably, and that lightness was of vital importance in a beam which was spanning 45 metres and penetrating into the space.

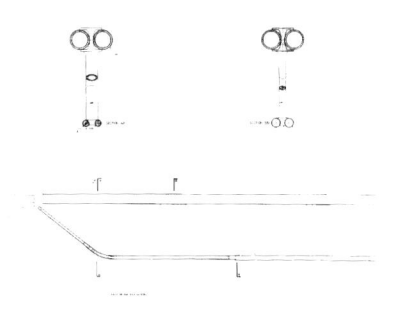

Sections and elevation of a double boom beam, Pompidou Centre, Paris.

The complete structural solution for the floor of the Lloyd's of London building came to my mind one evening in Turin while I was working on the Fiat project. The grid and the relationship between the columns and the floor beam grid were all there in the concept from the beginning. We had been discussing the problem very thoroughly during the previous weeks, so it is not surprising that I should have been thinking of a possible solution. But that the solution should arise fully formed is, I think, unusual. It is characteristic of the way engineers think: because they are working with objective parameters these lead to only one conclusion.

When I discussed this proposition with Richard Rogers he averred that it had always been true of the work of engineers. The bridges of Telford and Brunel, he felt, had probably arisen in their final form from the beginning and the engineers only needed to prove that they worked rather than modify and change them to get them exactly right. I think this is how I work. Once a solution appears to solve the problem, then I don't feel any desire or compulsion to change it. If it's a solution it is a solution and so be it.

I think this characteristic of working in concepts is quite common for engineers. It arises with architects too sometimes, but it is probably a fundamental aspect of all engineering designs.

Invention, innovation and creativity are three of the great buzz words of today. To be creative or innovative, especially, implies a God-given talent, possessed by the few for the rest of us to admire. From the outside it seems impossible to achieve these qualities, as though somehow the gifted appear to have some mystical status, a status that cannot be claimed, only conferred.

How can we become innovative or creative? As I have argued already, the creative talent is essentially artistic and is essentially associated with architects, designers or artists. For the engineers or other practitioners in science or fact-based information, the aim is to innovate. Is this so different? I think not.

It is a myth that there is something special about the innovative engineer. Probably every solution put forward by an engineer has some unusual element, some feature that could be called innovative, but is not recognized because it is buried in an otherwise conventional solution. And if we examine the nature of these otherwise innovative or inventive elements, we will find that it is just the result of the engineer being intelligent or sensible about the way some detail has always been, and so reassessing the problem from another point of view.

This kind of innovation arises a hundred times a day on building sites throughout the world. It is taken for granted. But it is nonetheless an important part of the everyday work of engineers everywhere. Every solution involves some original thought, some special contribution which we would classify as innovation. This need not be spectacular, it is enough to be new or original.

A spider's way of structure
Open-ended research leads to the most exciting results and stimulus. The spider and the structural engineer share similar requirements and constraints when designing a new structure. The key to the structure of the spider's web lies in its shape and stress distribution. By allowing large elongation of the threads, the maximum proportion of kinetic energy from a flying insect is absorbed as strain energy. The multiple redundancy of the radial threads ensures that the web will function even if many radials break.

A distinct structural hierarchy in the web is defined by the large difference in prestress. Radial threads are highly tensioned compared to the spirals. If a bee hits and sticks to a few strands of spiral thread they break the impact. The load travels from the spirals to the radials, where much of the kinetic energy is absorbed, and then continues straight out to the supports. Meanwhile, the whole web gently vibrates, dissipating energy through air resistance.

The spider is using the techniques of the late twentieth century engineer, but with much more elegance and precision.

Jean Prouvé

Jean Prouvé came into my life towards the end of the Centre Pompidou design and before construction began. His enthusiastic support was an important factor in our belief that we could get the steelwork built and erected as we had designed it, in spite of general opposition from the French engineering establishment and industry. That he should have supported us will surprise no-one. He had been a member of the competition jury and had made his support of our project well known. And he was hardly a member of the engineering establishment. His background and lack of formal engineering education had seen to that. Nonetheless his opinion counted. His support mattered. It mattered because he was one of the great natural engineers of the middle years of the twentieth century, untutored, unconventional, a maverick engineer and architectural inventor, whose understanding of materials and process was at once precise and unlimiting, enabling him to do things that a more conventional engineer would have found impossible.

Natural engineering talent is rare today. Public accountability and public responsibility require that everything be calculated, checked and endlessly analyzed by computers using the latest theories. In such a world the natural genius without formal education moves elsewhere, into boat design maybe, into motor-racing perhaps, somewhere where skill, talent and understanding matter more than proof and where proof can be achieved by performance. It was not always so. In the nineteenth century and before all the great structures were the work of natural engineers. Gradually their work and the rules which govern it were codified. And slowly that codification became more important than the original fountain from which it sprang. Society demanded that architecture and engineering should only be designed by people who were specially trained in these arts. Natural engineers and builders are being replaced. They have no place in our specialized society. This is sad, as much talent is thereby suppressed. People whose understanding of materials and how they should be used is instinctive and physical, as distinct from mathematical, are no longer able to survive in this climate. They play a minor role. Our society is much poorer for this development, for these 'improvements'. And Prouvé and his work provide the proof of this. He was the last, or almost the last, of a long line.

The role of the engineer in our society is to guarantee the integrity of all the artefacts of our lives. Whether concerned with a building,

Jean Prouvé, 1901–1984.

An Engineer Imagines

Maison Tropicale, 1949, under construction.

Maison Tropicale, cross-section.

Maison Tropicale at Niamey, 1949–51.

a bridge, a space rocket or a refrigerator, engineers have to guarantee that it will function correctly. They must analyze how it will perform under the different service conditions which can arise. This responsibility makes engineers conservative and dismissive of those who do not conform, or who do not understand the rigours of the role that they play. Generally when faced with a design problem, they proceed carefully, assessing and analysing the way a machine or structure will perform. They will check whether the materials are appropriate, whether the artefact can be built, whether it will be understood and therefore correctly used. All of these characteristics are evident in Prouvé's work. His attention to detail, his interest in manufacture and construction have all the hallmarks of a well-honed engineering approach.

But, unlike most engineers, for Jean Prouvé that was only the beginning. For him the necessary engineering virtues were taken for granted. They were the platform from which his inventiveness sprang. He was always inventing. Faced with a problem, his instinct was to invent a solution which synthesized all the conventional requirements with the engineering needs of constructability and performance. In doing this he was concerned with the appropriate use of materials, especially the then new materials of extruded aluminium and sheet steel.

Just as industrial engineers in the automobile and aircraft industries were beginning to understand how to exploit the lightness and versatility of these processes to make safer and lighter cars and aeroplanes, Prouvé was exploring their use in the building industry. His various essays in prefabricated, lightweight steel and aluminium structures are classic examples of how to exploit the engineering properties of sheet steel. His Tropical House is a beautiful clear example. At its core is a pressed metal portal frame, built on a locally provided base. This frame provides the horizontal stability for the whole house. On it the roof and walls are supported. This enables them to be kept light and simple. The shape of the central portal itself reflects the way it works, resisting the horizontal wind and perhaps earthquake load. By choosing pressed metal, this relatively complex geometry is easily achieved, and a slight curvature and corner reinforcement guarantees the maximum efficiency of performance of the otherwise fragile material. These were properties and characteristics then being explored and used by other industries.

Jean Prouvé was very up to date. But there is more. The outer

Jean Prouvé

Maison Tropicale. Assembly of prototype at Maxeville, 1949

light-reflective skin of the Tropical House is separated from the inner insulated skin. Natural cooling and ventilation are used. There is moveable shading to control sunlight and direct the ventilation where it is required. All the elements are flat for easy packing and transport and to give the minimum of fuss in assembly. This small, almost neglected, building would be a perfect answer in today's energy-conscious world, with its low consumption of energy in use and, perhaps more important, minimum use of energy-consuming materials in manufacture and construction. The building is full of subtle details which make a study of it a delight for a professional. The use of a separate suspended floor above the locally made base provides insulation and helps control damp. The shape and geometry of the ventilation chimney in the centre benefit from the stack effect, the tendency of hot air to rise, and, as with a fireplace, this can be used to control the amount of ventilation or natural cooling needed.

The balance between constructability, materials and function is pure engineering virtuosity. Prouvé's other tropical structures, like the Prospector's Hut, are the same. They are inventive and original, but firmly rooted in sound and clearly expressed principles.

Whichever structure of Prouvé's you examine, you find the same simple clarity filled with elegant engineering solutions. Two other particular examples which have always impressed and fascinated me are the Maison du Peuple at Clichy, built in 1938, and the façade of the CNIT in La Défense, Paris, built in 1959.

Maison Tropicale, detail of the adjustable *brise soleil*.

At Clichy the external panels are in lightweight steel sheet. To avoid the surface distortions caused by buckling, a spring is held inside, giving the sheets a slight outward curve. This enables a very lightweight panel to be used, but with consistent form and without unsightly surface distortion. Even today when many steel sheet panels are used, you do not find such simple elegant consistency, which stems from a clear engineering understanding of the materials being used.

Air transportation of the Maison Tropicale, 1949

At the CNIT the façade has many features designed to improve its performance. It is hung from above and horizontally braced by the walkways. These walkways give access for maintenance and cleaning, all of which can be carried out from the inside. The overlap of function of the horizontal passerelles between access and structure, the opening façade, which gives access for changing as well as providing ventilation, is clear engineering thought at its best, unclut-

An Engineer Imagines

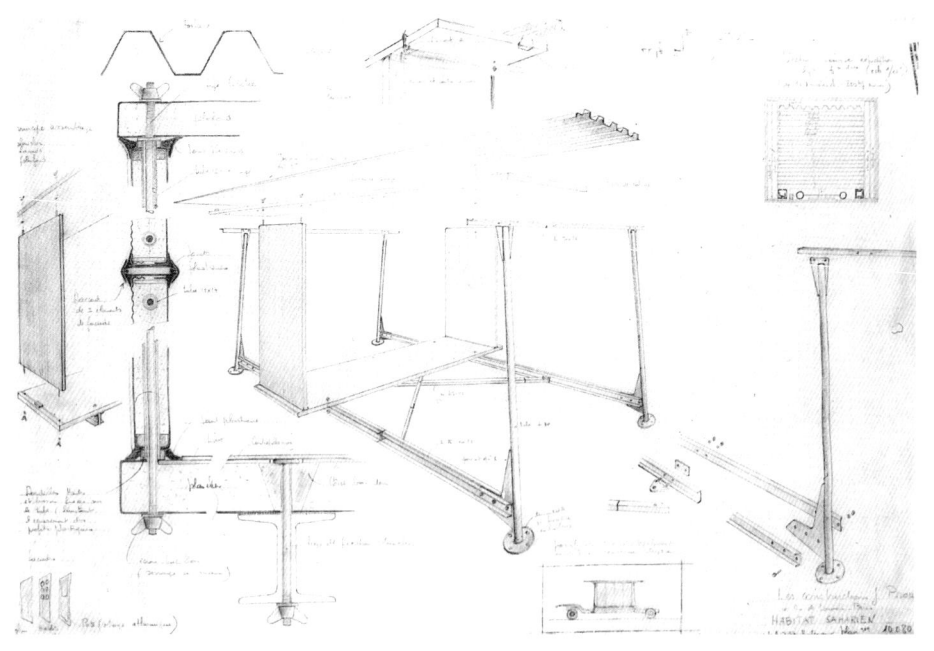

Maison pour le prospecteur solitaire du Sahara. Details of assembly and transportation, 1957.

Maison pour le prospecteur solitaire du Sahara, 1957.

Maison du Peuple at Clichy, 1938–1939.

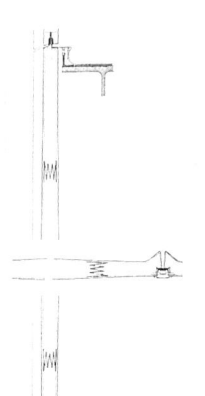

Maison du Peuple, skin detail. The springs used for stressing the metal sheets played a very important part in the final visual effect achieved by the façades at Clichy, since they equalized the surface planes and tended to impart a double convex shape to them.

Jean Prouvé

View of the CNIT, steel and glass façade by Prouvé, with new café terrace and entrance structure desgined by RFR.

tered by preconception, examining the real problem and creating a simple obvious solution.

Prouvé called himself (or maybe the word was employed by others seeking a label) a constructor, and I am writing of Prouvé as an engineer. Of course the two are intertwined, as they should be, but often they are not, especially today. Jean Prouvé lived at a time when the scale of construction and the industry which serves it was smaller. There were not the colossal companies doing everything, making all the decisions, where everything serves corporate policy.

He was an individual. He invented. He expressed himself. He liked to know how things worked, how they were made, how the materials which were in them could best be deployed, and to use that as the basis of his art. That is in essence the definition, the true and proper definition, of an engineer. Was Prouvé an engineer? Of course he was; he was the embodiment of what an engineer should be. He was a man of his time.

In our rigorous and controlled modern world, the world of collective responsibility, it would have been more difficult for Prouvé. In his work he anticipated the way we design and use materials today. Many of his designs are uncompromisingly modern, especially in the way he used and expressed materials. When you look at his structures, you see pure engineering simply expressed. Careful study is required because that simplicity does not come easily. Every bolt, every joint counts. Everything has a reason and that reason may be just as much in the requirements of the fabrication as in the need to withstand load.

Prouvé, with his comprehensive engineering approach, is an icon. The values that he represented must be preserved because they are

Detail of CNIT façade showing articulated mullion profiles with angled glass panes.

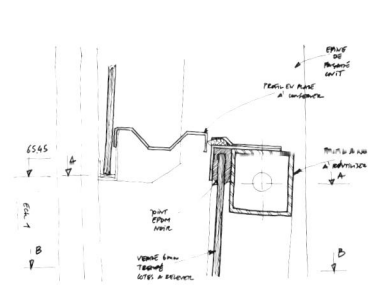

CNIT façade, section through transom.

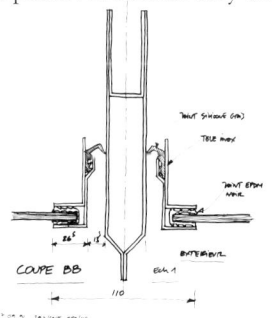

CNIT façade, section through mullion.

85

Menil Museum against the downtown Houston skyline.

Menil

The Menil Collection Gallery in Houston, Texas, was one of the most important projects in forming and in confirming my attitudes to materials. It is an exemplary project in many ways. It is primarily about light. Mrs De Menil, the client, had an outstanding art collection which had been assembled over many years. She had decided to leave the collection to the city of Houston, to be housed in a gallery which was funded by her and certain eminent members of the Houston community. She had decided from the beginning that the art should be seen in natural conditions. By this she meant that in daytime there would be a direct relationship between the levels of light inside the museum and the outside: that is, when the sun came out, the inside would light up with the extra external light.

Renzo Piano, who started the project shortly after the winding up of Atelier Piano & Rice, visited a variety of museums with Mrs De Menil. Together they decided that natural light transmitted through the roof plane would give the most direct contact between the gallery spaces and the outside. A louvre system was chosen for the roof.

During the time of Piano & Rice we had experimented with various materials. Ferro-cement and ductile iron were two of these materials. In combining them we sought to weave together the porcelain-like fragility of the ductile iron with the soft, grainy texture of the ferro-cement into a continuous melded whole.

Renzo, with active support from me, decided that the Menil gallery roof would be in ferro-cement. This is a very highly reinforced thin sheet of concrete. It was invented by Pier Luigi Nervi, the outstanding concrete designer of the age, in Italy in the 1950s.

The ferro-cement, which is usually made by a plastering technique and has seen its principal use in boats, is a material dependent on very high-quality workmanship to make it sound and durable. In its normal condition it is 1.5–4 centimetres in thickness with six to eight layers of wire mesh reinforcement. When it is used in boats, a steel cage is made to approximately the shape wanted. This cage, which has occasional longer rods to help keep its shape, becomes a rigid hull. Very dry mortar is then made and the cage is plastered and then finished on the outside by wooden and steel trowelling. It is very important for the quality of the final project that the plastering is done in a single continuous application and that no dry joints occur (that is, joints between mortar which has already started to harden and wet mortar which is being applied).

Cross-section through museum showing basement, main gallery level and treasure house where paintings are kept in darkness when not on display.

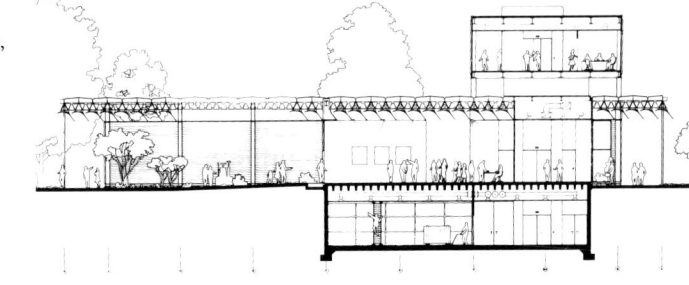

Sketches of the Menil Museum leaves by Peter Rice.

An early sketch by Renzo Piano integrating ferro-cement louvres into a space-frame structure.

Sketch of a Menil Museum leaf by Peter Rice.

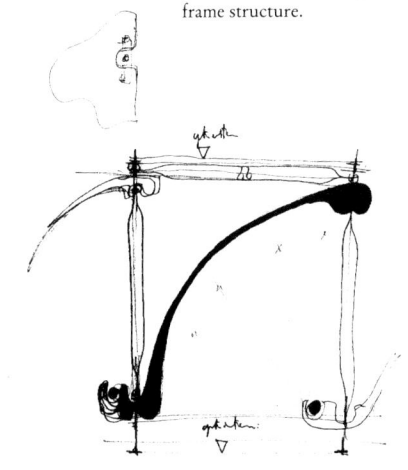

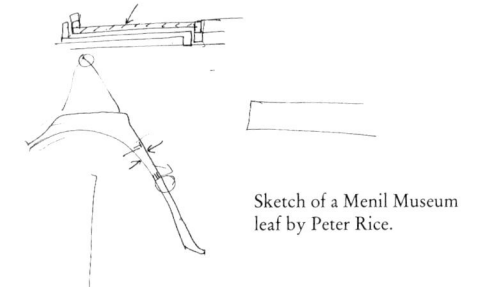

The essence of ferro-cement is that it has a very high toughness and uses the minimum amount of materials. It was invented for a situation where materials are expensive and labour costs are low. Low labour costs usually indicate that good-quality craft labour will also be available. Most of the recent development and research into the material has been done in India. But it is also a material which, because of its fineness, low thickness and plastered finish, is capable of very elegant shapes and surface quality.

The plastering is an essential feature of the manufacturing process. It guarantees a very dense matrix of mortar in the spaces between the mesh and the quality of that matrix can be inspected while finishing the outer surface. So, in deciding to use ferro-cement, we were choosing to use a craft-produced material in an industrial environment. Houston, we thought, being near the sea would have an important boat- and yacht-building tradition. We could be lucky.

When Renzo produced his first sketch design he showed a space frame with the ferro-cement louvres integrated into the structure as shear elements, replacing the diagonals. I quickly realized that this would not work, as it defined the form of the louvres before we knew the requirements for the light. I proposed instead that the louvres be separate elements, forming the lower members of a truss with the steel elements above to complete the structure of the roof. The stage was set for detailed development of the roof to begin.

At this time the person doing the work on light was Tom Barker at Arups. Tom, together with Mrs De Menil and Paul Winkler of the Menil Foundation, developed a strategy for the conservation of the paintings and the amount of light to be admitted. The roof had three layers. The outer glass surface was the waterproof surface. This was a conventional glass roof, in so far as a horizontal water barrier, in an area of tropical rainfall like Houston, could be considered conventional. The glass in this reflected out a high proportion of the heat and the ultra-violet light which will damage paintings. In the space between the glass and the ferro-cement were the air-conditioning ducts.

The first thing to do was to find what shape to make the louvres. As the USA is a well ordered country – the street system runs east west, north south – the louvres were easy to align with the sun directions. The shape which developed slowly was a product both of the shape Renzo wanted and the need to cut out the sunlight. The strat-

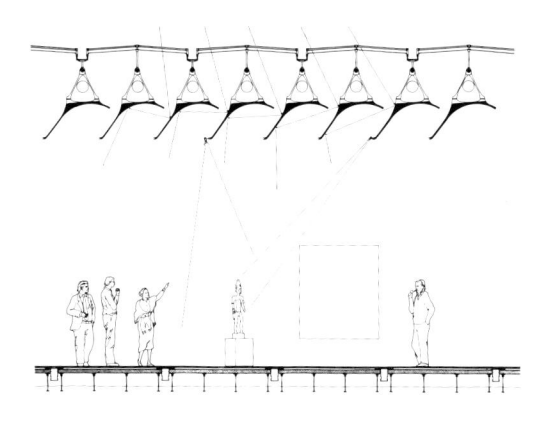

Section through a typical gallery, showing entry and internal reflection of natural light through ferro-cement leaves.

egy developed for conservation was to estimate the total amount of daylight and artificial light which would fall on any piece of art. Provided this total, mathematically integrated over a year, was no more than would occur from normal conservation levels constantly for the year, then the paintings would not be unduly damaged. The paintings were assumed to be on display for one month out of six, or six months out of thirty-six. The rest of the time they would be stored under ideal conditions in a specially constructed storage house, where they could be viewed briefly by scholars and people with a special interest. Thus the levels of light permissible during their time on display was on average six times that which would normally be permitted. This gave a range of light levels which could then be allowed in the gallery. Direct sunlight was of course excluded. An upper band of 2,000 lux was chosen to define the maximum light that a work of art would ever suffer. The general aim was to create an inside ambience as close as possible to the actual conditions in which the paintings were made. When outside light conditions were very high (in sunlight the outside light would reach 120,000 lux), then the inside light would reach 2,000 lux. In darker outside conditions the gallery would be darker in proportion. Thus lighting in the gallery exactly reflected the light outside. When a cloud passed in front of the sun, a cloud passed over the gallery at the same time. The natural connection was achieved.

Once we had chosen the shape of the louvres – for some reason we called them leaves – we had to decide how we might make them. We were certain that we could not tolerate the geometrical inaccuracies which the traditional process would have introduced. To check this and to verify that the computer methods that we were using to calculate the light levels were correct, we constructed a mock-up, or prototype gallery space. With this we calibrated the computer programme and then used data on the Houston weather recorded over a number of years. In this way we could check both the maximum and minimum light and the integrated illumination exposure the gallery would see during a typical year. This enabled us to define accurately the dimensions of the leaf. It is a measure of the sensitivity of the shape to error that the presence of the rail for the artificial lighting at the tail of the leaf made a big difference to the results. Everything had to be included.

Having decided that we could not accept the geometrical errors which the traditional process would bring, we looked for an alter-

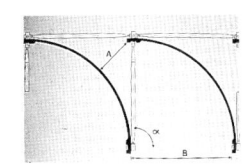

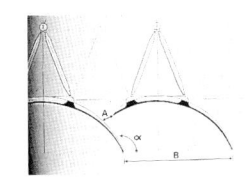

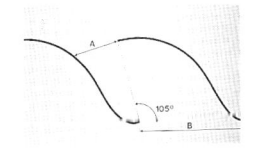

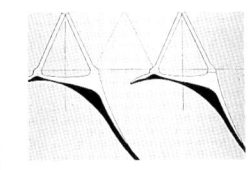

Evolution of the design profile of the ferro-cement leaves determined through a combination of lighting tests, structural requirements and architectural sculpting.

Peter Rice, Renzo Piano, and Mrs de Menil.

Early prototype study.

Renzo Piano and Peter Rice examining a prototype leaf.

native method of manufacture. We discovered an engineer in California who had a patent for making ferro-cement using a spray concrete technique. This patent had never been exploited. At first we tried to persuade precast concrete companies in the Houston area to be interested in the process. None was. We searched for boatbuilders who might be able to manufacture the leaves by the plastering process but found none. Eventually we found a company in England which had set out to exploit this patent and we decided to work with them.

We made a fibreglass mould of a leaf and then made a back-up mould to act as a surface to spray on to; we developed a way to manufacture the multi-layers of mesh to place into the mould. The key problem was how to guarantee that the mortar matrix was dense. Obviously, if it came through the steel layers in a dense form and looked solid and sound on the formed surface, that was a good indication. Early checks convinced us, however, that large flat water pockets could develop parallel to the mesh layers as the force of the spraying caused vibration of the mesh. The process was modified and some extra steps were introduced. With these extra steps and with careful monitoring of the process, we were able to guarantee the quality of the final product. It is interesting to think about what this meant. In most of industry the quality of product is guaranteed by testing after manufacture. Here was an element which we could not test after it was made. The guarantee of its integrity lay in the process of making it. So we were relying on the process and monitoring the process to ensure we had the quality we needed.

One had to be sure that a level of commitment to quality existed. The final inside surface was created by plasterers' trowelling, giving the quality and geometrical guarantees we needed. The sand we used was marble sand and the cement was white, which gave a sparkle to the finish of the surface. The manufacture of the leaves was a central feature in the quality of the project, and without the use of this patent it is doubtful if we could have achieved the quality and accuracy required.

The second part of the project was the trussing structure mentioned above. Here I decided that we should use ductile iron. Ductile iron is a form of iron which does not have the usual brittleness of cast iron. It was invented in Britain at the end of the 1940s. Cast iron is different from cast steel, the casting material used in Centre Pompidou. Ductile iron gets ductility because the carbon crystals

Fabrication of shell element

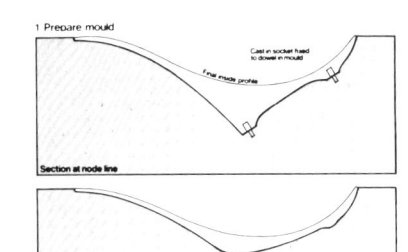

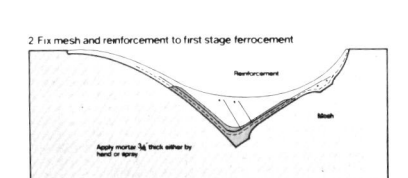

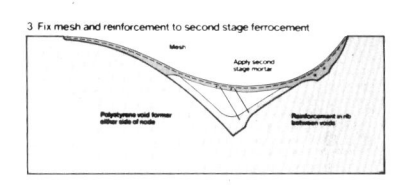

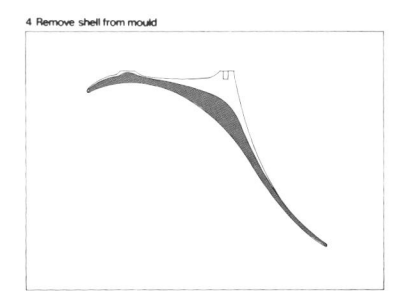

1 Prepare mould

Section at node line

Section between nodes

2 Fix mesh and reinforcement to first stage ferrocement

3 Fix mesh and reinforcement to second stage ferrocement

4 Remove shell from mould

Fabrication of ferro-cement leaves.

Ferro-cement being shot into mould.

Hand-finishing final surface of leaf.

Detail of ferro-cement leaves.

Working session in Genoa. Left to right: Shunji Ishida, Peter Rice, W Paul Kelly, Tom Barker and Renzo Piano with a ductile iron prototype leaf truss section in foreground.

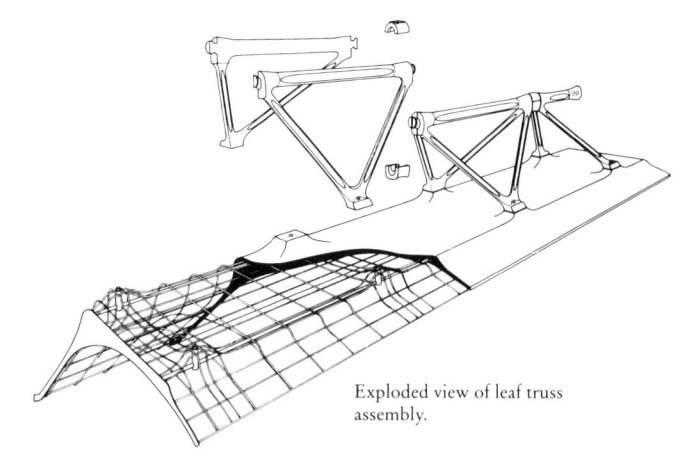

Exploded view of leaf truss assembly.

Assembling early prototype of ductile iron truss.

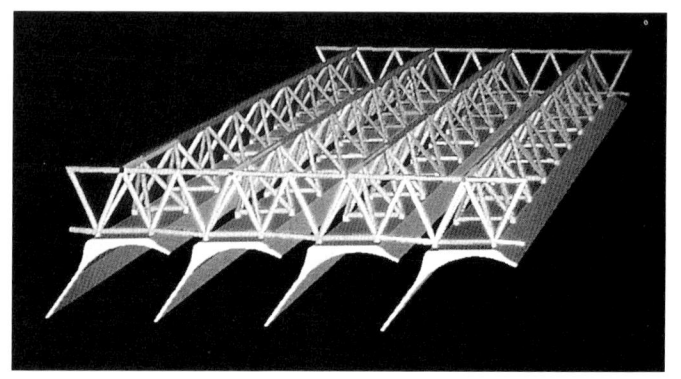

Computer model of four leaf trusses supported by girder trusses at either end with leaves hanging below. This was used to confirm effective shading of gallery from direct sun.

Trial assembly of first girder truss castings.

Bolted girder connection detail.

Checking bolted connections in truss.

Site inspection of roof truss.

coagulate into round balls and do not have the sharp surfaces of normal cast iron. I had long wanted to use ductile cast iron, another idea carried from my past.

During the work on Centre Pompidou we toyed with the possibility of using cast ductile iron for the main glazing mullions where they would have been a visually lighter alternative than the trusses actually used. We did not have the confidence to make them in this way; they were rather large and would have been difficult to cast. Cast iron differs from cast steel in being much more fluid while it is being poured during manufacture. This means that it can be made into much finer and more fragile shapes. Thus the intended elegant match for the ferro-cement could be more easily achieved.

There was one further advantage. Cast steel during its manufacture must be heat-treated twice, once just after casting and again later after it has been examined for flaws and any found have been repaired. This causes distortion of the piece, which in turn means that all points of contact between adjacent pieces must be machined. In Centre Pompidou we made a feature of this and used the fact to help provide control of future additions to the building. Here we wanted to explore a different approach, where we would make a continuous clamped iron frame whose nature and lack of bolted joints would be a fragile complement to the ferro-cement. Ductile cast iron has a form where it does not need to be heat-treated after casting and therefore the mould controls the geometry and good dimensional control can be achieved. This in turn meant that a

Detail showing complete roof section from external glazing above to ferro-cement leaves below.

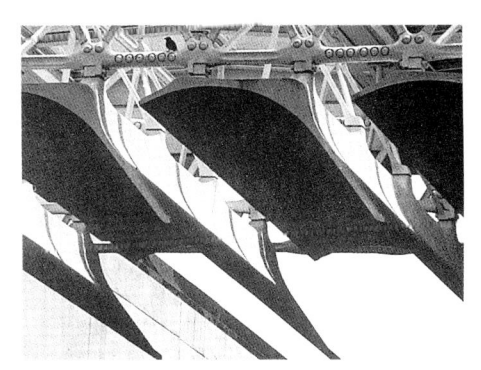

Detail of leaves.

Interior view of Contemporary Gallery at west end of Museum.

clamp detail can be used at the joint between ductile iron members.

Together these advantages justify the use of cast iron on this project. The detail of the clamp and the smooth transition between the iron and the ferro-cement express the different physical characteristics of the materials used and differentiate them from other similar materials, such as precast concrete, which might have been used. They serve the general objective of giving physical presence to the materials of structure so as to materialize the physical nature and working of the structure.

The special quality of the light and contact between the internal spaces of the gallery and the outside weather add a diurnal quality to the experience of visiting the gallery. The control of light and predicting the condition within the gallery is now something which is possible using modern computer methods and could not have been done at earlier times.

Fabric

Lightweight structures is a broad, if rather inappropriate, name for a group of surface structures made from fabrics or tension or compression nets. The structures are lightweight because nets or fabric are the lightest available materials with adequate structural properties to span in two directions. Tents and nets have, of course, been in existence for a long time. Recently, however, stimulated by the early work of Frei Otto and others, a new understanding and vocabulary of possible shapes has emerged.

Originally the research attempted to define the factors which made for efficient structural forms. In doing so it examined structures in nature, to see how they solved the relationship between form and efficiency. The most interesting and important work stemmed from a study of soap film surfaces, forms which are generated by the characteristics of the soap film itself, a uniform surface tension. Modelling techniques to generate soap film surfaces were developed. Modelwork led to an appreciation of the transient nature of the soap film surface; the surface form is dependent upon the load conditions acting upon it. It also led to the realization that modelling techniques generally were the best way to study and understand the relationship between geometry and equilibrium for these surfaces. The modelling techniques brought with them too an indication of the way in which real structures could be created. It is an essential part of the study of lightweight structures that the forms, the initial load conditions and the method of manufacture are wholly interrelated.

Later, computer methods, based on Finite Element Analysis, were developed to mimic the modelling process. These techniques together have now reached the stage where the limits of what can be designed and built are the limitations of materials and the limitations of the designer's inventiveness, not, as has hitherto been the case, the limitation of analysis and specification methods.

When we talk of these structures we are talking of free forms, which must be in equilibrium with themselves and with a set of applied forces. Usually they are very light so that, in the initial condition, gravity loads are not very important. However, to counteract non-uniform applied loads, either snow- or wind-induced, they are almost always prestressed, either by air pressure or directly through tensioning of the net and fabric elements. To understand the physical characteristics of any particular lightweight form it is still important to work with models. One has to learn through expe-

Spider's web: a natural cable net.

Study for 'Arctic City', architect, Frei Otto. An air-supported roof dome spanning 2km to enclose a town of 40,000 inhabitants in controlled climate.

From a soap film surface study by Frei Otto at the Institute for Lightweight Structures, University of Stuttgart, Germany.

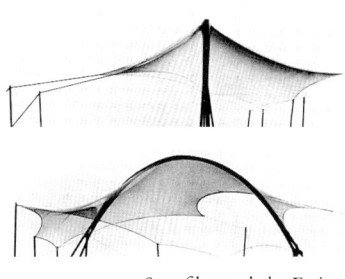

Soap film study by Frei Otto for Kuwait Sports City.

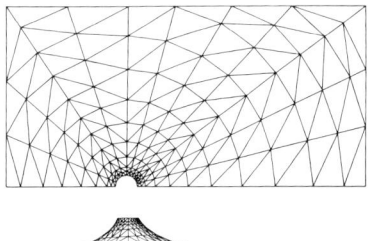

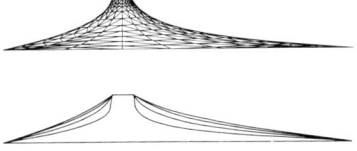

A 'finite element mesh' is the computer's version of the soap film. The membrane curvature can be altered by changing the balance of tension between one direction and another.

rience the vital relationship between stiffness, geometry or curvature and the stability of the surface under changing loads. The modelling is used to help generate surfaces which have a balanced relationship between curvature in one direction and curvature in the other and between curvature at different parts of the surface.

Fabric is a highly specialized subject. To design something in fabric, you have to guarantee that the shape has certain characteristics. The principal characteristic it must have is an anti-clastic curvature at all points in the surface. The term 'anti-clastic' means that if you take two cross-sections at right angles to each other, the curvature in one direction is the opposite to the curvature in the other direction, so that the fabric is tensioned against itself. This means that if you want to design something in fabric, you have to generate a form which has these characteristics, otherwise the fabric will flap in windy conditions or under other kinds of non-regular loading and it will tend to destroy itself.

There are a whole variety of ways in which you can achieve this anti-clastic curvature in terms of the form. You can introduce simple tent-like structures, or you can introduce arches or other kinds of structural elements into the surface to create one particular line of curvature which is then used to offset the other curvatures. Certain areas of the fabric can be flat provided there is a sufficient amount of prestress in the surface. That's the second characteristic that you have to have. To prestress a fabric means that you have to be able to tension it up in such a way that under no-load the fabric has got tension in both directions throughout its surface. And this tension is such that one direction is pulling against the other direction of tension. It's possible to have flat areas of fabric provided they're not too extensive and provided they're bounded by areas of fabric which are not flat.

There are different kinds of fabric. But most of the kind of structures I have been designing use Teflon-coated glass fibre (polytetrafluoroethylene; PTFE), which is a relatively stiff fabric, permanent in nature. We are talking about a permanent building material because the Teflon is a non-deteriorating plastic and the glass fibre that is the load-carrying element within the fabric has the characteristics of glass so it is comparatively stiff. This means that high levels of accuracy and precision are needed when defining the fabric shape so that at no point in the surface is there any lack of prestress and at no point in the surface do you get creasing. But all of

Schlumberger Headquarters, Montrouge, France. The anti-clastic surface is formed by flying masts which support the roof membrane from beneath, while tension is introduced by pulling the membrane outwards against the two side buildings. Note longitudinal seaming.

this is a specialist technology which is quite well known.

The most important thing about fabric is that it's normally a translucent material. The way the fabric is assembled to create the final form and your perception of its shape, normally seen from the inside, is given to you by the seaming which arises from the cutting patterns. The joining together of those pieces to create the perfect surface means that the curvature and shape of the fabric itself are difficult to see, particularly from the outside and particularly with PTFE fabric, which is pure white once it has bleached to its equilibrium condition. It's only really the doubling up of the jointing, the shadows of the seams which you get at the points where the cutting pattern panels meet, that enables you to perceive the surface form. This is a very important consideration. It means great care must be given to the selection of cutting patterns, so that the perception of the overall surface that you ultimately see is consistent with and enhanced by the detailing.

To give an example of the importance of that, I would cite two particular projects which I worked on a number of years ago. One is the Schlumberger Montrouge Tent in Paris for the Schlumberger garden entrance area. It's quite a small tent, about 15 metres by 100 metres, where the cutting patterns were made in such a way that they formed continuous strips longitudinally down the length of the tent. The minimum amount of cross-patterns ensures that the sense of longitudinal continuity is maintained.

Another project of a similar size where the same approach was used is in the Thomson factory research facility at Conflans Ste Honorine outside Paris. Here the fabric is supported by a series of arches. The fabric is patterned to provide longitudinal continuity in the same way as the Schlumberger tent, so that you don't get an interruption to the way in which you look down the space.

Many of the designs which we have created have been relatively simple at that level. Another one is the fabric roof at Bari, where the patterning is also reduced to minimize the interruption to viewing. Yet another example is the tent at Lord's Cricket Ground in London. For reasons of cost, this is made of PVC with a special external coating. Here too the whole of the perception of the space and the way in which the fabric is used to create the internal ambience come from the way the patterning has been done: the architect has used radial patterns to emphasize or create a sense of individual space underneath each canopy.

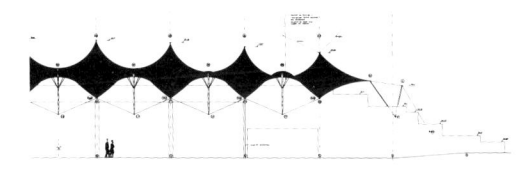

Interior view of Bari Sports Stadium, Italy, showing the subtle combination of a slender rib-frame with membrane.

Lord's Mound Stand,
London. Interior view
showing radial seaming.

Fabric

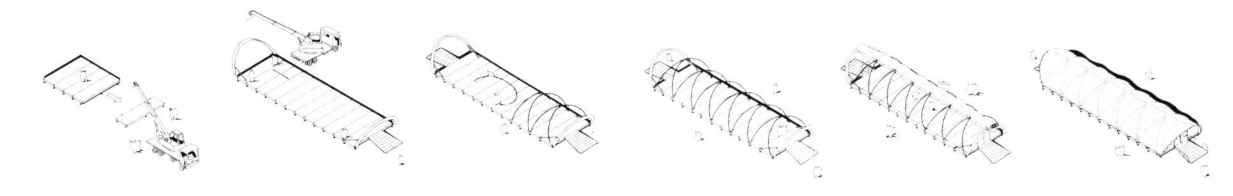

Erection sequence of demountable MOMI Tent.

When the architects, Future Systems, received the commission for a demountable structure to be used on London's South Bank for receptions for film festivals and other special occasions, it was evident that we would have to design something as light and ephemeral as possible. Future Systems, and especially Jan Kaplicky, have over the years uncompromisingly supported and promoted the use of the latest technology in design.

Tenara, woven and laminated, is a very light Teflon-based fabric and, as it is PTFE, it has the property of being self-cleaning. It also has a much higher translucency than any other Teflon-based fabric. We chose a cylindrical form for the South Bank structure, employing pultrusions, the technology of fishing rods, to enable the use of special visually light details.

These arches were laterally supported by the fabric and did not have a continuous inside member. We felt that to have a member inside the arch would create a virtual surface when you looked down the surface on entry. This is an idea which was also explored in Centre d'Art Contemporain in Luxembourg by the architect I M Pei. By having a non-continuous inside member all the emphasis is transferred to the outside member and the fabric surface. The use of the glass-fibre rods, 3.2 centimetres diameter, for the arches gives the public a direct relationship with something they understand. Therefore the sense and feeling of lightness comes not just from the size of the arch member but also from the material of fabrication, for a fishing rod is an element understood by everyone and is known to be light and flexible.

It is to be hoped that this combination of a fragile light fabric and a visually discontinuous light structure explores the meaning of a lightweight demountable structure. The ends, which retain the fabric in tension, are made of a steel frame with transparent sheet stretched across it. This material, although it appears fragile and prone to damage, is in fact very tough, especially when stretched.

So in general terms I think fabric, once you've accepted the special nature of the technology that you have to use, and once you have accepted the discipline of understanding the patterning, is a relatively straightforward design problem. Sometimes however I feel that fabric which, if you like, is a sheet, doesn't have much presence. The kind of presence it has, say, in the Schlumberger project in Paris is about the limit of what you can achieve seen from the outside and there it's really providing a continuity across the chasm.

Section through MOMI Tent showing pultruded rib which is braced in plane by flying struts and cables. Single-piece membrane is pulled down over ribs and tensioned against floor.

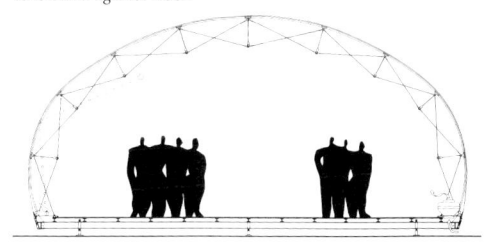

Interior view of MOMI Tent showing anti-clastic surface the membrane takes up between ribs. Transparent Hostaflon plastic is used to enclose ends.

An Engineer Imagines

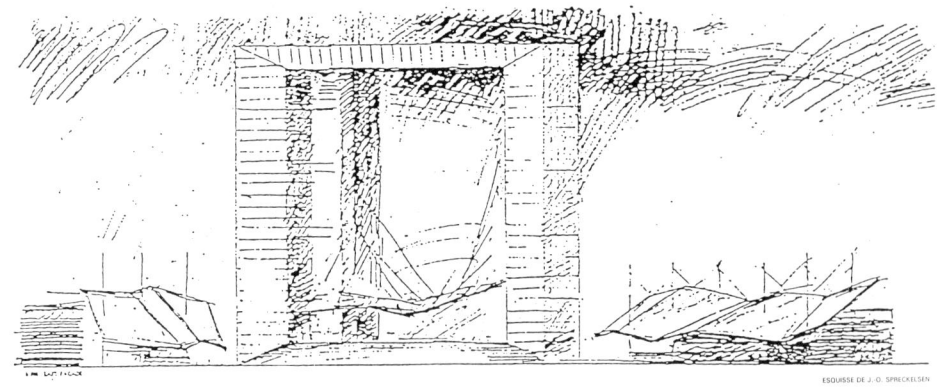

Spreckelsen's competition sketch for the Nuage Cube.

Nuage with clouds.

View of the Nuage inside La Grande Arche.

It's really a single external roof providing continuity between the two sides with the variety coming from the mix of form which is used to provide the shape of the tent itself.

One of the things that is quite important to recognize, particularly if one is using Teflon-coated glass fibre but also if you're using some of the specially coated PVC solutions, is that the external surface is highly reflective. Teflon as a coating material, apart from being non-stick and therefore self-cleaning, starts off yellow but in time becomes white. It's highly reflective and there's nothing one can do about that, but it means that the ability to see the nature of the surface from the outside is limited.

For some time I have felt that, as an object in space, fabric has very little physical presence and I've been looking for ways in which one can give physical context to a fabric surface. A recent opportunity to look at that came with the Nuages for the Cube (now called the Grande Arche) at La Défense, Paris. That project arose because I was introduced to J O Spreckelsen, the architect for the project, shortly before he resigned. The design of the Cube project was the subject of a major international competition that Spreckelsen won. In the design for the Cube Arch itself, he had conceived an internal structure; its purpose was to give scale and measure to an immense volume. He had originally intended it to be a series of glass planes but for a variety of reasons that proved to be impossible. The structure was too heavy and the whole thing became out of scale. He asked me if I would work with him to develop a fabric alternative, but before we could get very far in the development of that he had resigned and I was left working with Paul Andreu, who was the executive architect for the project.

Spreckelsen's original fabric intention was one large sheet-like structure inside the space. I felt that while this might provide a certain amount of shelter in the space, it was very unlikely to provide any perception of scale, so I sought to think of the solution in a different way. I was worried initially about what I would call the presence-of-the-fabric solution, because I felt that a single-sheet solution would tend to disappear in certain circumstances to do with light and viewing angle and would not provide any of what might be called physical body with which to work. These things were called Nuages or clouds and in this context I felt that the thing had to have body, depth. It had to occupy a certain physical space.

The other thing that was important in this particular project was that it was in a very exposed position. It was on an east–west axis, elevated above the surrounding topography so the wind loads and the wind conditions under which the fabric would have to work were extreme. The shape of the arch itself created a kind of wind-tunnel effect. I felt that if we were going to design something which would both give presence and offer a sense of scale, it was very important to detail it in such a way that the key element in it was not the fabric but some structure supporting the fabric. And with that in mind I developed, working directly with Andreu, the context of a solution which said that the Nuage was not just a fabric sheet but was a fabric sheet supported by a set of cable trusses which together represented a physical volumetric presence similar to that of a real cloud. Thus it was the combined presence of the structure plus fabric which became the Nuage. The real volume occupied by the fabric was joined by the virtual volume of the steel structure.

Instead of trying to make the fabric span large distances, the fabric part of the Nuage itself was broken down into sails. I introduced the idea of glass into these sails so that somebody underneath could see through the glass to the arch above; therefore, although enclosed under the translucent fabric canopy, you would also be able to see through it in places to get a sense of the grandeur of the arch itself.

Again here the patterning of individual sails, the cable trusses which are joined to a cable boundary and the detailing necessary for the erection became, as with other projects I've worked on, the definition of the scale of the project. In other words, it's the detail which you see close up. The individual sails are then patterned to emphasize the glass panel in such a way that your eyes are drawn to look up into the arch space above.

In the end the project was quite complex, because the arch itself had already been designed and was actually half built by the time we became involved; this determined the size of the forces that we could apply to the suspended elements. It was important also that, as far as possible, there was no support from below. It was necessary to tie the canopy down, but beyond that we didn't want to have any vertical support below. We wanted it to be entirely suspended from above so that the sense of freedom of the element suspended in space was not damaged or destroyed. This feeling of a floating canopy with the cable trusses providing a physical presence and therefore a physical depth to the whole thing became the basis of the idea.

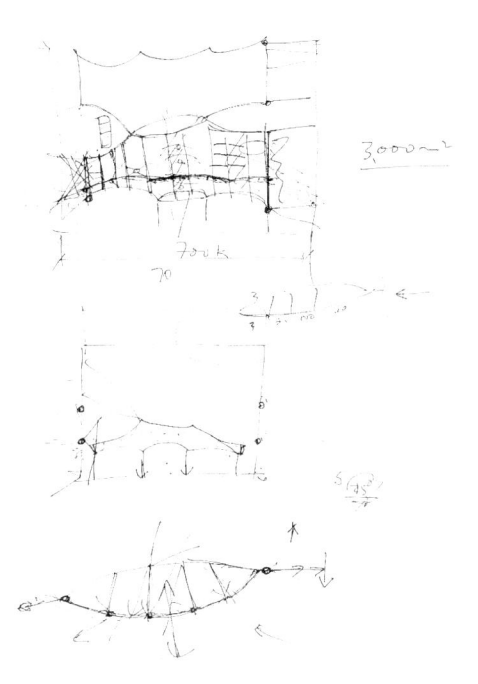

Early sketches dicussing Nuage structure.

Nuage Cube. Principle suspension point between the two canopies. The membrane is stressed against the cable trusses using flying struts. The surface is punctuated by glass discs.

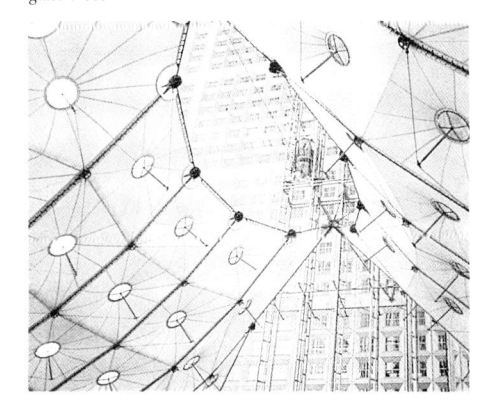

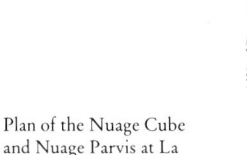

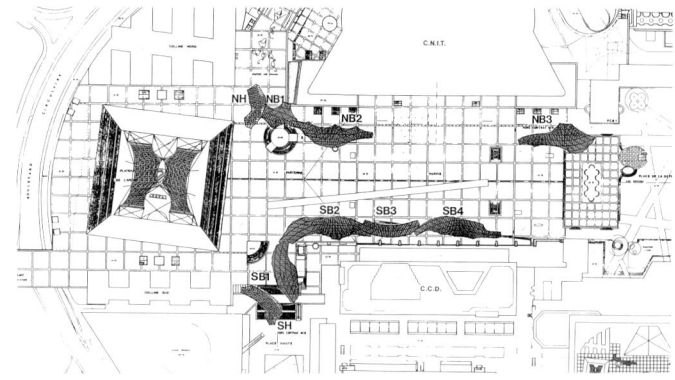

An Engineer Imagines

Plan of the Nuage Cube and Nuage Parvis at La Défense, Paris.

Model of Nuage Parvis showing how the different cloud canopies might intersect and overlap, varying light and spatial qualities below.

Computer model of Nuage Parvis in front of La Grande Arche.

Interestingly, this was something which people afterwards found very difficult to understand. The project didn't get very good reviews. It was perceived as heavy, which was not a product of the design concept so much as a product of the extreme wind conditions that the fabric had to withstand. But it was not perceived as light and airy as it should have been at the time when it was first built. I think it's become much more accepted since then. As it has mellowed and as it has become more accepted as part of the Cube and its environs, I think it has become much better understood. People are now, I feel, beginning to recognize the nature and character of what it is and that the details, the cables, the fabric, the glass, the richness of the whole actually enhance the scale of the arch.

When Spreckelsen originally conceived the Nuage, he proposed not just a Nuage in the Cube but that a Nuage should come floating down in front of the Cube in the space on either side. These were called the Nuages Parvis, the parvis being the platform in front of the Cube. There was a problem here because the client for the Cube was not the same as the client for the parvis. The client for the Cube was the SAEM, a special government organization set up to build the Cube and its Nuage and a couple of buildings on either side of the Cube. But the part in front of the Cube was the responsibility of the *Etablissement Publique* for La Défense, EPAD. Therefore it was not possible to design the Nuages for the parvis at the same time as the Nuage for the Cube.

The Nuages for the parvis, which were designed later but not built, produced an approach to fabric design that I think represents an advance in what we have managed to do to date. The idea of the Nuages Parvis was important in terms of the cost of fabric. It's very important that you get repetition. If you don't, you have to utilize separate cutting patterns for each and every panel and the cost of providing cutting patterns and cutting the fabric separately is a substantial part of the total cost of the finished structure. But we were also looking for something which had a free form, something which wasn't rigidly defined by a regular geometry.

What we did was to examine how, using current computer technology, you could design an individual sail or individual section of the fabric bounded by one curved free geometry in such a way that the standard panel was not rectangular. It had a non-regular shape but it was still possible to repeat it so that the left-hand boundary of the panel could be joined to the right-hand boundary of the adja-

Fabric

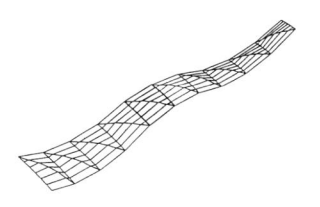

cent one, and so on. Using this standard panel, one could create a non-uniform free-form geometry. In order to obtain the fluctuations of the Clouds, this module was part of a skew surface (neither flat nor orthogonal) limited by sinusoidal curves.

It's only possible to do this using current high-quality computer geometries using a process analogous to the method of constructing fractal images. What we're looking at is the idea of using the computer to generate a geometry from a standard panel which itself is a complex shape, but out of which, by exploring a variety of assemblies in the computer, one could get a total shape that to all intents and purposes is free-form but made from standard panels. And I think that as a matter of principle, using the computer as a creative design element – not just as a replacement of something that one would be able to do by hand, but actually doing something which you couldn't otherwise have done – represents a whole new set of possibilities that can be explored with standard fabric panels.

Two or three projects have been designed using this technique, but none of them has yet been realized and it's unlikely that the Nuages Parvis will now be built because there's a substantial body of opinion within French architectural circles which says that to put something there would ruin the sense of grandeur of the space. I think there's also a lot of opposition from the people whose buildings would be partly obscured if the Nuages were there.

This approach is only valid when you're thinking of the fabric as an object, not as a cover, whereas in projects such as Lord's and Bari the fabric is used essentially as a cover seen from below. However, in Bari the fabric's presence is part of the total form of the stadium and the way in which it is integrated into the other elements of the stadium means that its physical presence is guaranteed. But, the fabric itself is entirely viewed from below and therefore the cutting pattern approach is quite valid.

I feel the approach we used for the Nuages is valid because it's a way of exploring these fabrics. What's interesting is that the kind of shapes you can generate with fabric couldn't be generated with glass or with other rigid materials. You have to have this essential geometric freedom. The properties fabric offers give you the opportunity to explore these free forms. And the thing that's interesting about this is that once you postulate a panel type with the possibility of it being joined together in more than one type of configuration, you don't actually know what the outcome will be until

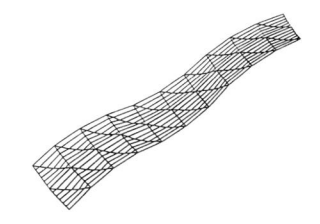

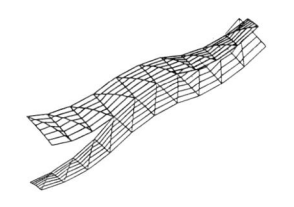

Study for the Nuage Parvis showing the variety of surfaces which can be built up from a standard element.

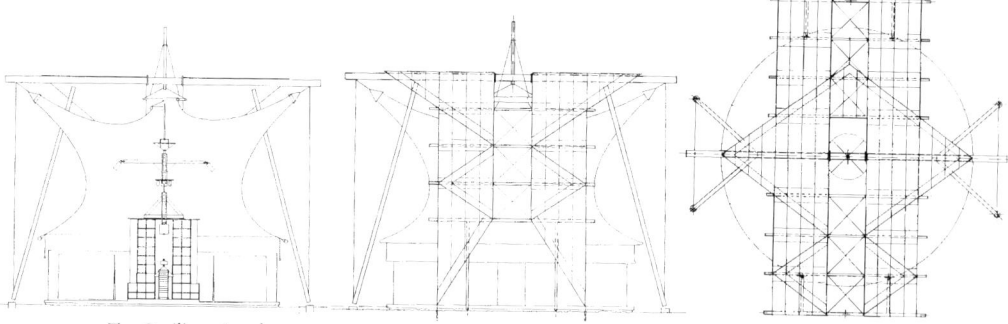

Tate Pavilions, London.
Elevation, section and plan.

Tate Pavilions, London.
Erection sequence.

Tate Pavilions, London.
Designers' references: early
nineteenth century bed.

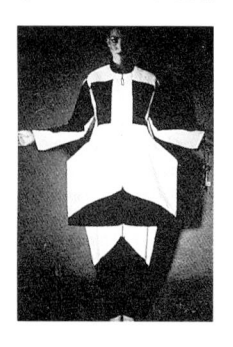

Tate Pavilions, London.
Designers' references:
Stepanova dress, c. 1921.

you've explored it on the computer. Therefore you're not so much predesigning something that you want to build as postulating an approach within which you then identify a key element. And then you have to explore it on the computer to see what it gives you. I find the idea that the process is an integral part of the design quite interesting, rather than seeking to achieve something that is a pre-determined paper form.

Of course there are lots of other ways in which fabric is used that aren't structural, taut if you like. There are all sorts of other areas of life, such as dress-designing, where people explore fabric and the way in which it can be created to give very special forms and shapes. I think that there are a lot of additional areas using fabric which we can explore. It's interesting that when we were looking at the Nanyang project for Singapore, we used as a point of departure a Kandinsky painting. The particular characteristic of that painting that we were interested in was the way in which different layers of surface overlapped each other to create different densities of light. It was a loose point of departure rather than a particular attempt to follow a particular painting. Not 'loose' in terms of its application but more the philosophical intent of what one might do. It was more the idea that by overlapping fabric, in some cases you would be under three layers of fabric, in others you'd be under two layers. The density of light underneath and the sense of space above you would be informed, not just by a single fabric surface but by over-lapping fabric layers in much the same way as Kandinsky explores the three-dimensional nature of planes in some of his paintings. I think there are others as well that you could use for that kind of inspiration.

There is also the potential of mixing different translucencies and transparencies. One of the things that I think has been inadequately explored is the mix of fabric and glass or polycarbonate, because fabric of itself produces an almost shadow-free light. If you have a fabric roof over a space, then you get a completely diffused light coming through it, visually flattening its curvature, and I think that the actual nature of the space inside, a roof or something like that, can be a little lost. So one of the other things we've been doing has been looking at or examining or trying to design structures where you have a mix of translucent and transparent panels. That's what we did on the Nuage Cube, but there is also another project, the Bull Gambetta project in Paris covering a large light well inside the

Fabric

Wassily Kandinsky,
1866–1944. Untitled, 1925.

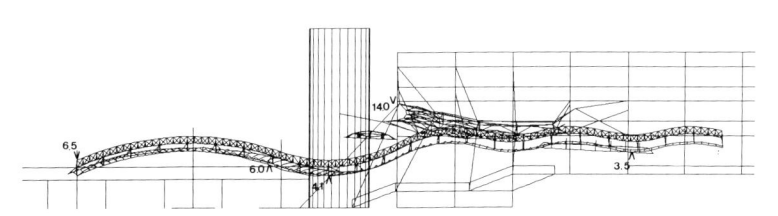

Nanyang, Singapore,
section.

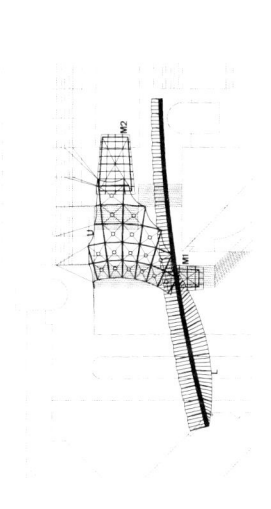

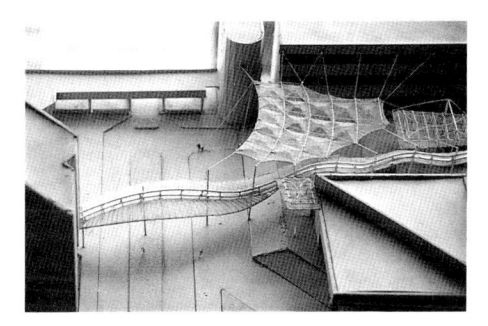

Nanyang, Singapore,
model.

Nanyang, Singapore, plan.

Bull Offices, Avenue
Gambetta, Paris. Atrium
roof combines fabric and
polycarbonate panels.

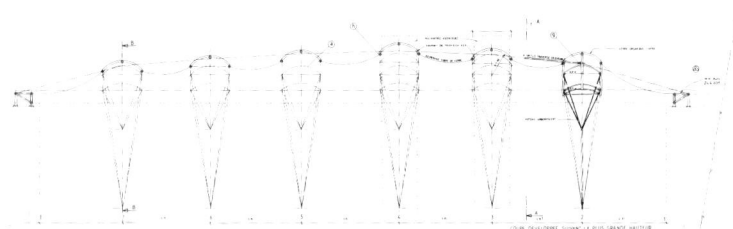

Bull Offices, Avenue
Gambetta, Paris.
Longitudinal section
through atrium.

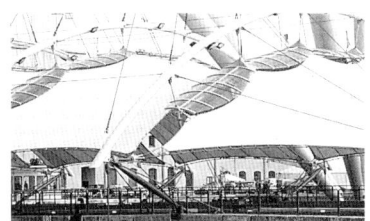

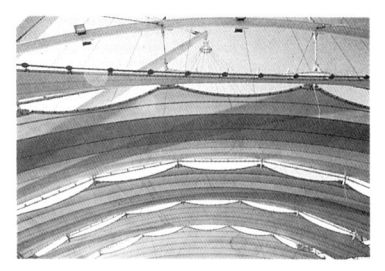

An Engineer Imagines

building, where we had a mix of fabric and polycarbonate to provide a more animated roof form.

One of the problems in a way is that fabric is often used as a cheap solution. This offers very little scope for manœuvre. But in situations where one is looking to achieve a certain architectural effect, then features such as the overlap in the canopies or the introduction of translucent and transparent panels, together with the growing number of fabric types offering different levels of translucency that are going to be available, mean that it'll become increasingly possible to explore different mixes of translucency and transparency, qualities of light and perception.

Il Grande Bigo, Expo '92, Genoa, Italy. Suspended fabric canopy with transparent glass panels.

Lighting mast with fabric 'bird' light diffuser, Esch, Luxembourg.

Esch at night.

Glass and Polycarbonate

At the beginning of the 1980s, I designed two projects at the same time which helped clarify for me the way in which the physical characteristics of a material can be used to influence the way in which the detailing and form of the solution are developed. They were the Grandes Serres at La Villette and the IBM Travelling Pavilion. Both were projects where the essential reason for using the principal material was its transparency. In La Villette we used glass and for the IBM Pavilion, polycarbonate and timber.

In both cases we were hoping that the transparency of the main material would lead to a feeling of 'non-materiality' and, although separated from the structure, the public would feel a sense of communication through the transparent skin. We wanted both the serres and the pavilion to be light and seemingly fragile. The choice of glass and polycarbonate was easily made.

Polycarbonate has a clear lightness which means that the sense of presence of the material is ephemeral and permits one to see images of objects without any colour distortion or lack of clarity. The IBM project was undertaken with Renzo Piano; he had defined it by making a romantic prototype which he photographed and showed to the client to demonstrate the central idea. The polycarbonate was used because it is robust. The pavilion was scheduled to be transported to some twenty major European cities and re-erected each time in a city-centre site. Obviously it had to be demountable and the detailing was chosen to facilitate this.

The concept was to have a series of polycarbonate pyramids which, together with timber inner and outer members, would act as a semi-circular arch spanning twelve metres and creating an internal space which would, when placed in a suitable natural setting, give the sense of being with nature. The purpose of the exhibition was to present IBM and its computers to young people as natural tools which would enhance life and our future enjoyment, as opposed to the commonly held perception of them as alien to our natural development.

In his original proposal Renzo had combined the timber with cast aluminium to emphasize the natural quality of the space. Polycarbonate as a material is light and robust but it is also not very strong: its load-carrying capacity and its stiffness are low. These were the principal characteristics which we sought to express in the design. Making the polycarbonate perform a structural role meant that all the joints had to perform properly and predictably.

IBM Pavilion, prototype.

An Engineer Imagines

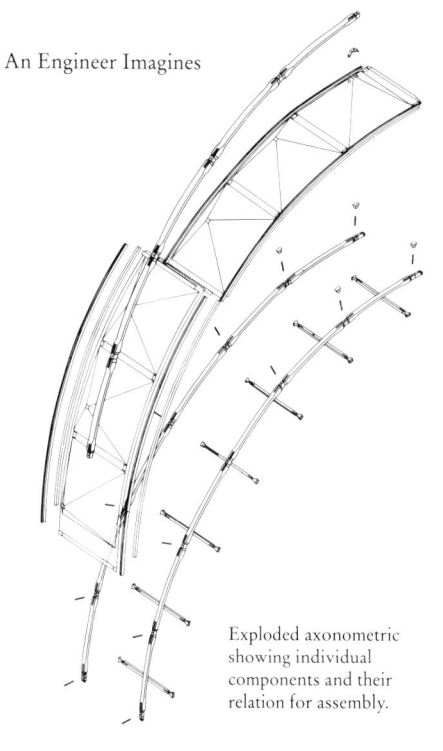

Half of pavilion cross-section showing separation of timber chord from polycarbonate.

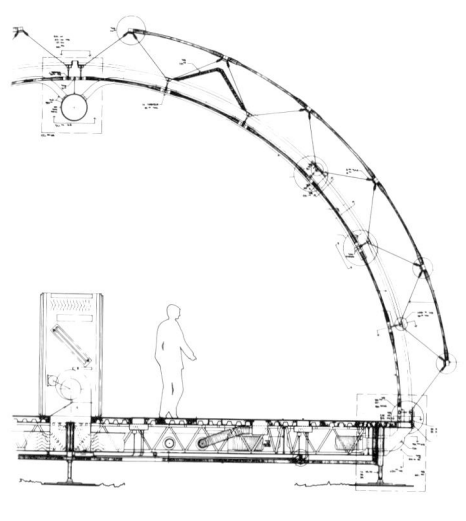

Exploded axonometric showing individual components and their relation for assembly.

Transparency of polycarbonate connects inside of pavilion to nature outside.

Renzo Piano, Peter Rice and Neil Noble reviewing components from original prototype in Renzo's garden at Pegli, Italy.

Interior view of IBM computer exhibition.

Stacks of polycarbonate pyramids waiting to be unloaded from truck.

Two halves of an arch being assembled on the ground.

The arch halves are then fixed to the floor structure, jacked up into position where they are pinned together at the top to form a three-pin arch.

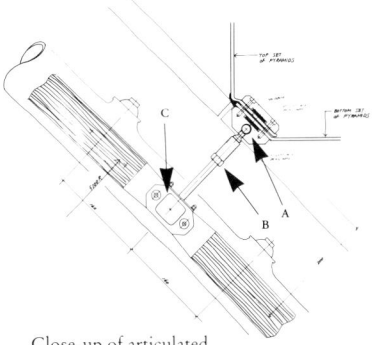

Glass and Polycarbonate

Close-up of articulated strut showing the ability for adjusting the length and indicating its pinned fixing to the polycarbonate.
A Rotational fixing to pyramid.
B Adjustment for length.
C Rubber block cast on end of strut enables push-fit into casting and allows rotation.

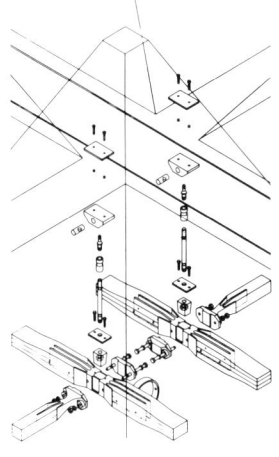

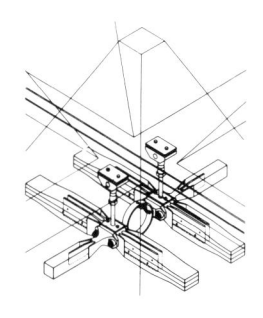

Inner chord of polycarbonate pyramid: stages of assembly on-site and final assembly.

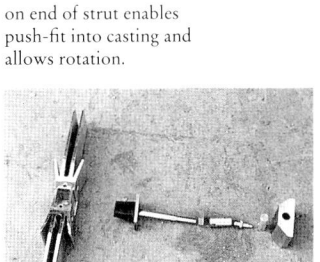

Components for articulated strut connection.

Typical internal connection showing relationship between inner timber ladder and the pyramids.

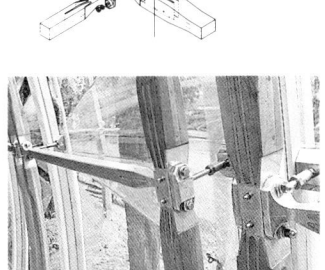

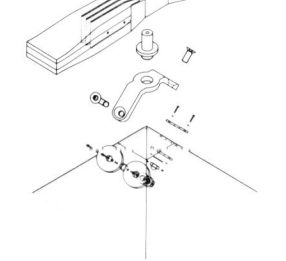

Typical pinned connection at springing point of timber ladder.

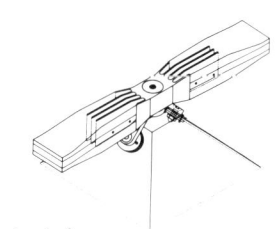

Outer chord of polycarbonate pyramid: stages of assembly on-site and final assembly.

Components that make junction between crown of pyramid and outer timber chord. Rubber block allows push-fit connection to be made on site.

Connection between outer timber chord and crown of pyramid.

Overall exterior view of Pavilion.

109

An Engineer Imagines

Grandes Serres, Cité des Sciences et de l'Industrie, La Villette.

The serres (or conservatory structures), being glass, had almost the opposite material character. Conceived by the architect Adrien Fainsilber, they were a visual transition zone between the museum space of the building and the park which surrounded the museum. The park was designed by Bernard Tschumi after an international competition. However the serres remained and the definition of their role as a transparent transition zone was also left unchanged. Transparency was the quality the design was expressing.

For a surface to be transparent the presence of the surface must be clearly defined. This and the other ideas we explored in the design of the serres are fully explained in my earlier book *Le Verre Structurel*. It is sufficient to say here that glass as a material is very different from polycarbonate. It is very strong but fragile. To be fragile means that it breaks easily, particularly under shock or sharp loading. All this we know. Glass is after all a material with which we are all familiar.

We were seeking to build a plane of glass, which from the general dimensions of the serre would be eight metres by eight metres. The maximum size of manufacture of the glass we wanted to use was two metres by two metres. We decided to adopt a suspended assembly of sixteen sheets two metres by two metres each to make the total panel of eight metres by eight metres. To suspend the glass we realized that we should make use of the great strength of glass when it is loaded correctly in its plane. Each pane is suspended by its corners from the pane above it, up to a maximum of four panes in height. (Other projects have suspended up to eight panes, or in this case a 24-metre strip of glass.)

We decided to transfer the load through the glass by placing a small spherical bearing in the plane of the glass itself. This bearing guaranteed that the glass was always loaded in a predictable way and that we could calculate quite accurately what was the maximum load which might need to be transferred. The small surface of contact between the bearing and the glass exploits the high local strength of the glass. In addition, by choosing a highly articulated connection we could test its behaviour and be able to guarantee its performance in use. Even though the diameter of the bearing was only 32 millimetres and the glass only 12 millimetres in thickness, the capacity of this bearing was approximately four tonnes, many times the maximum load it would be expected to carry in use. The freedom to move guaranteed that we had a predictable combina-

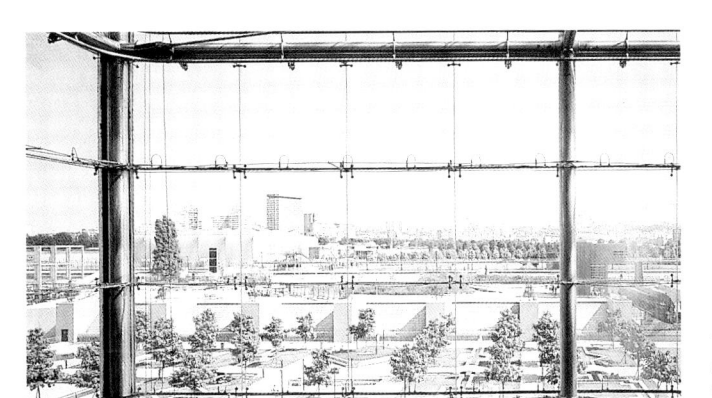

Glass and Polycarbonate

View through Grande Serre demonstrating transparency.

tion of load and strength of the assembly. Predictability is a very important quality of an engineering design. Furthermore the presence of the bearing in the plane of the glass defined that plane, from the inside and the outside. It offered minimum interference to vision, thus aiding the transparency of the assembly.

Prestressed cable trusses were chosen for the mullions to restrain the glass and were placed horizontally so as not to obstruct the panoramic view of the park seen from inside. Because glass is stiff and rigid in its own plane we were able to use the glass plane to restrain the mullion. We thus had the satisfactory situation where the glass was restrained by the cable truss mullion against wind loading, but the glass itself restrained the truss under load. This was possible because if the glass were not there to restrain the truss, then there would be no need for the mullion since there would be nothing loading it.

When we compare this detail from the glass with the equivalent details from the polycarbonate we can see how the properties of the materials have dominated the way the design choices were made.

Polycarbonate, because it is not very strong, needs a large surface of contact between it and any member transferring load to it. This is translated by a block of cast stainless steel which is glued directly on to the surface. It was possible to do this because the polycarbonate is not adversely affected by the uneven loading which the joint might transfer.

Polycarbonate moves a lot when it is heated. It is highly susceptible to the heat of the sun. Therefore the circumferential movement of the polycarbonate and the inner timber ladder had to be separated so that they could move independently of each other. This was achieved by placing a small strut between the two. The strut also enabled us to design a push-fit assembly joint between the polycarbonate and the timber to allow for a variety of conditions which might arise during on-site assembly. The push fit was made of cast-on rubber similar in size and hardness to an engine mounting. This permits a small amount of play in the joint, which prevents damage in handling and through incorrect alignment during assembly. The equivalent detail on the glass façade occurred at the point of support of the whole eight metres by eight metres panel. We inserted a connection to the main frame that uses a prestressed spring, which acts as a shock absorber for situations where unforeseen load changes occur. This could occur, for example, if panes

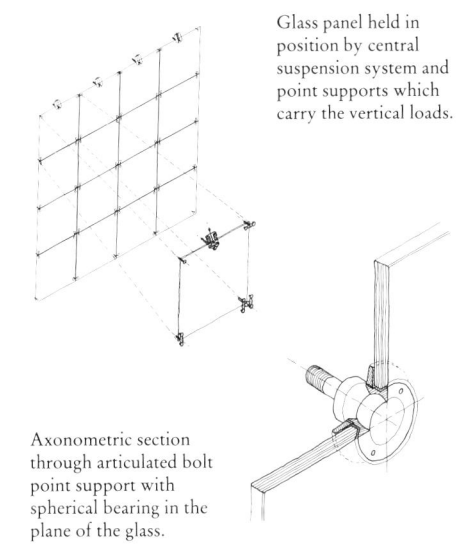

Glass panel held in position by central suspension system and point supports which carry the vertical loads.

Axonometric section through articulated bolt point support with spherical bearing in the plane of the glass.

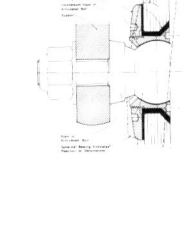

Section through point support.

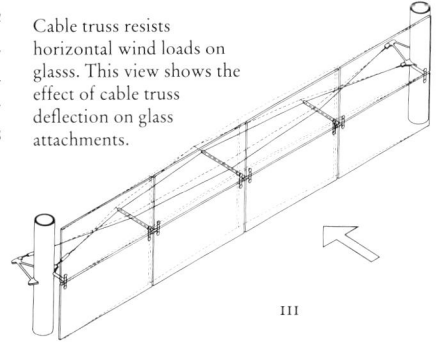

Cable truss resists horizontal wind loads on glasss. This view shows the effect of cable truss deflection on glass attachments.

III

Diagram of spring support.

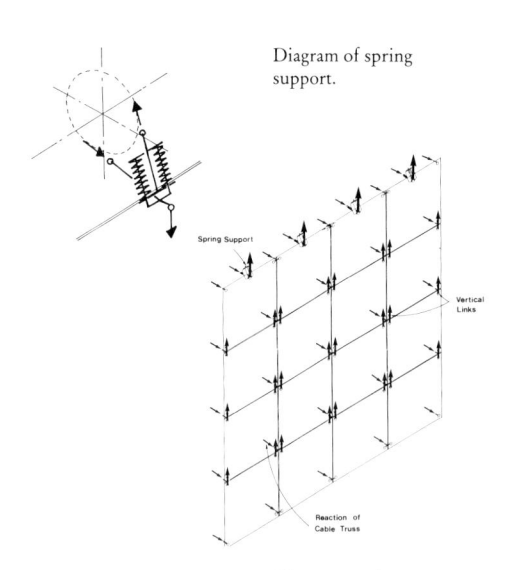

Glass suspension system.

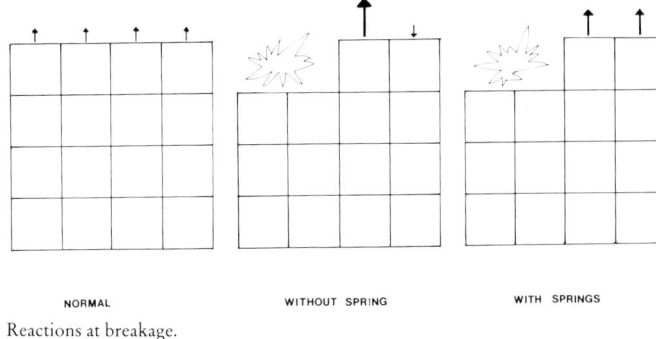

NORMAL   WITHOUT SPRING   WITH SPRINGS

Reactions at breakage.

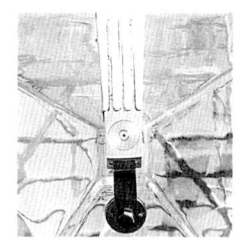

Exterior timber and aluminium finger connection at pyramid crown.

Interior view showing corner detail of cable trusses.

were broken in a panel. The redistribution of load this would create, taking place under shock conditions, could have overloaded an individual bearing assembly momentarily. The prestressed spring prevents this. It remains a stiff rigid support under normal working conditions, but under sudden overload conditions it takes the role of a fuse compressing under the increased load and thereby allowing the suspended panel assembly to readjust to the new load condition.

These two sets of details are a reflection of the physical difference between polycarbonate and glass. Other aspects of the two projects are worth noting. The cast aluminium connections in the IBM Pavilion offer a simple way of making the geometrical transition between the timber and the polycarbonate pyramids. The finger connection into the timber reflects the fact that timber is strong in direct stress and weak in shear. It is not strictly necessary in this situation as the forces are generally low, given the actual strength of the timber, but it is of symbolic importance given the attitude we have taken elsewhere. It also reflects the fact that the timber is a very workable material and can easily be cut to get the shapes needed to make the connection in this way.

We live today in an age of greater certainty. Everything can be calculated and is, and indeed must be, subjected to public scrutiny for safety, something which often leads to pedestrian and chunky designs. Legal liability, safety and the preformed prejudices of industry can be like a scythe which cut off the feet of any attempt at the seemingly bizarre or outrageous.

In the mid-nineteenth century materials were explored by building and waiting. Some of the structures built in this way, and even some of the great medieval glass windows, the rose window at York Minster for instance, would be difficult to justify today, with all our modern analysis techniques and our need to satisfy modern rules and norms on safety and soundness.

Does that mean that we are forever condemned to forms and shapes that are self-evidently safe and sound? This is one of the great challenges for the structural engineer and one which I have always wanted to meet head on. Obviously we must satisfy all these modern rules and regulations, which, if intelligently interpreted, leave a lot of scope for invention and innovation. 'Intelligently interpreted' – two sweet little words – dropped into the argument from nowhere. Intelligent interpretation is required not only by the

Rose window, York Minster.

designer but also by the checking authority, those who will give permission for a structure and assembly to be built. How often have I heard disbelief when I have explained the La Villette glass solution to a control engineer. 'That would not be possible here …!' is the response, before the engineer has examined the argument, and with the visible evidence of a completed structure to go and see. As if gravity, or the thickness of the window, could somehow change as you cross the border from France into Germany, or Switzerland or wherever.

It is no accident of time that both the La Villette and IBM projects first appeared in France where there exist the most intelligent and knowledgeable checking authorities that I have come across. The large centralized controlling offices, *bureaux de contrôle*, Socotec, Veritas, CEP and others each have at their head engineers who are equal in ability and understanding to any I have encountered in the best design offices, as Centre Pompidou amply demonstrated.

Peter Rice load testing a cable truss, Grand Serres, La Villette, Paris.

Lloyd's of London west elevation showing concrete 'structural diagram'.

End elevation of the Pompidou Centre showing steel 'structural diagram'.

Details: Steel at Beaubourg, Concrete at Lloyd's

Beaubourg and Lloyd's of London are two sides of the same coin. In one, Beaubourg, the structure is made in steel; in the other, Lloyd's, it is in concrete. The form and shape of the buildings are very different but the search for the solution had the same roots: scale and grain to articulate the face dominated the way the ideas developed. Concrete is a monolithic material. Steel is articulated. On Beaubourg we have an external steel structure providing scale and grain for the façade. In Lloyd's a similar solution is attempted in concrete. Lloyd's was designed by Richard Rogers about ten years after he worked with Renzo Piano on Beaubourg.

In Beaubourg the joints are articulated to ensure that we can be confident of the way in which the loads are being transferred. The gerberette is assembled with its machined surfaces in contact with other parts of the structure in a way that guarantees how the load is transferred between the main floor trusses and the column. It is very important to ensure that the load on the column is as close to its central axis as possible, such as to avoid any excess bending moment or twisting being transferred into the column. An uncentred load is referred to as an eccentric load. The column diameter and all the details which are dependent on it are designed to eliminate any eccentricity in the application of the load. Large spherical bearings guarantee this and ensure that we can predict the path that the load will take between the gerberette and the column. This assembly and all of the visible details which go with it are essential to keep the column diameter to its slender 800 millimetres. The 800-millimetre diameter is also a limitation imposed by the centrifugal casting fabrication method.

In this technique a cylindrical mould is spun around its axis as the molten steel is poured in — the more steel is put in the thicker the tube. In the Beaubourg project we wanted to use the same external diameter over the whole height of the column so that all the connection details would remain the same. As the load is small at the top and very large at the bottom a centrifugally cast column was an ideal solution, because the strength can be increased by increasing the wall thickness. This allows an optimization of thickness, which helps to contribute to the overall slender appearance that the columns have in the context of the building.

The floor is a steel composite of different members independently assembled. It delivers its load to the supports at the end of the large main beams. This means that it is easy to devise a way to

Interior view of Lloyd's atrium showing concrete columns and brackets which support the floor plates.

Beaubourg under construction showing steel gerberettes clasped around columns.

An Engineer Imagines

Structure of Beaubourg: steel elements are hierarchically arranged according to load path.

Column and gerberette: exploded view of components.

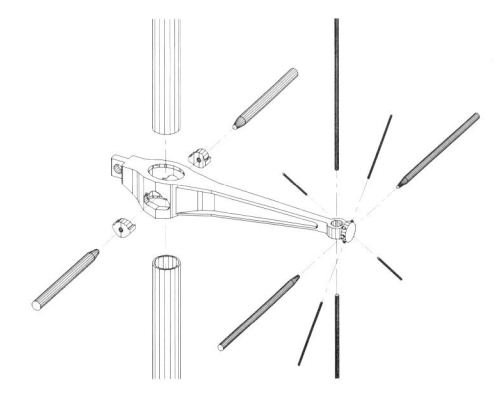

Column and gerberette under construction showing centrifugally spun column tube.

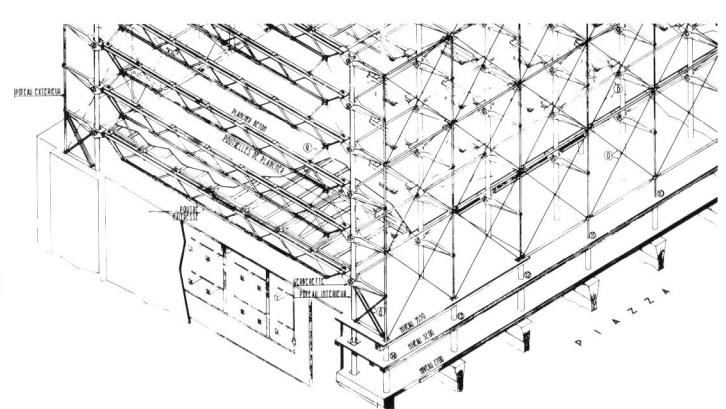

transfer this load to the columns through individual members. They – the primary and secondary beams, elements with specific articulated connections – are hierarchically arranged according to the load path. A steel building is an assembly of elements, the sizes of which are determined by the process of manufacture and the method of transport.

In concrete almost the reverse is true. Concrete is a cast material. This means that there is by the nature of construction a continuity between members when they are cast in place. When we decided that the Lloyd's building would be built in concrete, I realized that the expression of the joints and the separation of components so necessary to Richard Rogers' approach would be difficult to achieve.

It has sometimes been said that Lloyd's is a steel building built in concrete. It is true that at first we intended to make the external structure of steel, but when we changed to concrete there was no attempt simply to translate the earlier work. Our aim was to exploit the natural qualities of concrete while trying to achieve the visual articulation and legibility normally associated with steel. The project is an exploration of the material using both cast-in-place and precast elements while not rejecting other materials where these offered advantages. Structural steel connections were used for the precast elements of the satellite towers because precise, full-strength connections could be made quickly. Concrete-cased steel tubes were used for the main building bracing to minimize the diameter of this, essential to allow the space for bridges from the satellite service towers. Steel panels were used as permanent formwork for the main building floor slabs, because it was easier and quicker to build and provide a matrix of supports for the building services suspended below.

The cast-in-place floor grid was inspired by the Yale University Art Gallery, designed by Louis Kahn. What the architect wanted was a visible grid of beams which would unify the whole space. It was considered architecturally imperative that the Lloyd's Room should be homogenous and the same as any other office. A uniform grid of beams was the obvious solution. This grid could then be picked up from any point around its perimeter. The support columns were spaced at 10.8-metre centres around the perimeter. The plan shape is a rectangular doughnut with the central space used as an atrium, giving visual contact to all parts of the building.

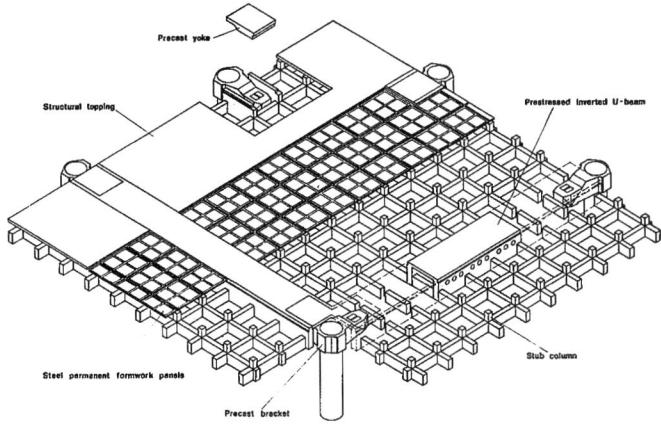

Axonometric diagram showing layering of Lloyd's floor system. Floor plate rests independently on brackets.

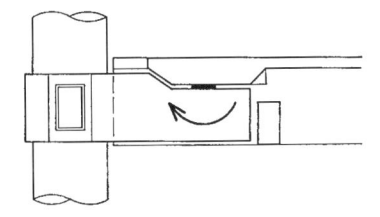

Because it is cast in place, the critical problem in concrete construction is quality control. It is vital to avoid joints, particularly those whose performance cannot be monitored in the formwork supporting the concrete in its fluid condition. Joints in the formwork or moulds are always a problem because the fluid concrete can leak during construction and spoil the nature of the finished work. We try to minimize the number and design them to reduce the risks. The formwork systems for the beam grid were developed to ensure the quality of the joints, while the joints themselves were detailed to help break up the concrete visually and reflect the method of construction.

The joints were also a key factor in the design of the column and its bracket, which is the equivalent of the column and gerberette of Beaubourg. The column was designed with brackets which extended like hands into the building to support the floors. The joint between the column and bracket was a good example of a place where the formwork joint was very sensitive. For reasons concerned with the function of the bracket, to achieve the required quality of appearance and to control this joint we decided to precast the bracket.

The bracket was cast in the form of a hollow cylinder with the 'hand' projecting. The cylinder was supported in position on top of the column which had already been cast and the cylinder filled with concrete to make the connection. The precision which could be achieved in the circular column and the precast bracket allowed us to create a joint with little risk of leakage.

Lloyd's: (TOP) load transfer from floor to column through bracket. Articulation between floor plate and bracket bearing allows rotation between floor and bracket. (ABOVE) bracket assembly during construction.

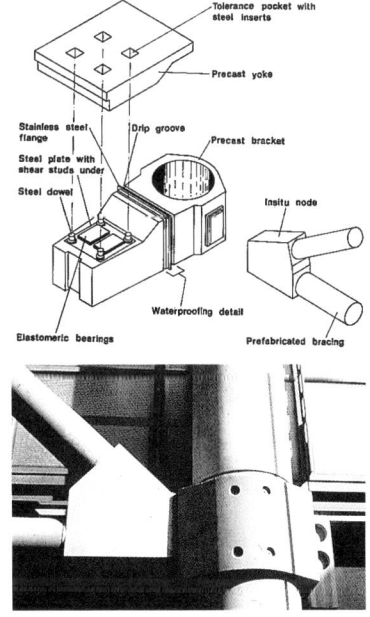

The concrete in the column has no difficulty in carrying the bending moment due to the bracket. Indeed, because it carries steel reinforcement, introduced to resist the tension which can develop, one can selectively add steel in such a way that the strength is enhanced. This is analogous to increasing the wall thickness of the Beaubourg columns. The weight of the building itself, transmitted down the column, also helps counteract the tension created by bending from the brackets.

In this way the bracket joint of Lloyd's is a direct reflection of the nature of concrete construction and is carrying out essentially the same job as the gerberette and bearing in steel used at Beaubourg – a case of two materials essentially doing the same thing.

Lloyd's: bracket, yoke, bearing and bracing; finished assembly (atypical condition).

Stone façade structure of
the Pavilion of the Future
during construction.

Stone: Pavilion of the Future, Seville

In 1988 I was asked by the Anglo-Spanish architect, David Mackay, a partner of Martorell Bohigas Mackay (MBM) in Barcelona, if I would participate with them in the design for the Pavilion of the Future. The site on the Isla de la Cartuja was 25 metres wide and 300 metres long, facing the old city across the Guadalquivir River, and alongside the gardens where many of Expo's night-time events would take place. The client, the Expo '92 Committee, specifically asked that the solution be spectacular.

I went to Barcelona to work with the architects and while I was there I remembered the Palacio da Ajuda in Lisbon, a building which I had seen the previous summer. This building, built around a quadrangle, had been left uncompleted on one side. Apparently Napoleon had attacked Lisbon while the building was under construction, and the work was stopped and never completed. A tall gaunt façade, complete with window openings but with no windows or building behind, was left to form one side of the quadrangle. I mused at the time that it was surprising that it stood up, but it was there, proof positive that it worked. It was not unlike the medieval ruins of churches, visible throughout Europe. Obviously such structures must be stable. I was very interested and thought one day I would build a structure like that. As I contemplated what to do for Seville, I remembered Lisbon and I proposed that we build a wall, a façade on to the park and separating the open space from the jumble of the international pavilions behind. The architect was interested, so we proceeded.

At first the idea took a form that derived from its inspiration. We thought to build an elaborate long façade as though there were a building behind. On examination that seemed complex to justify and very expensive. So we searched again. Then I remembered the experience of building the La Villette glass conservatories. We had postulated then that the same techniques might be used to build in stone. Stone and glass have similar physical characteristics, and we realized that the techniques developed for glass could also be developed to enable stone to be used structurally. Stone, like glass, is very strong in compression, but fragile and prone to cracking. If we could protect the stone from tension forces and from sudden loads then we could perhaps build the screen using stonework as a primary structural material, but in a more sophisticated way.

The possibility of working in stone led me to think of what else was being built in stone today. It was clear. The architects and

Palacio da Ajuda, Lisbon, Portugal. View of free-standing façade.

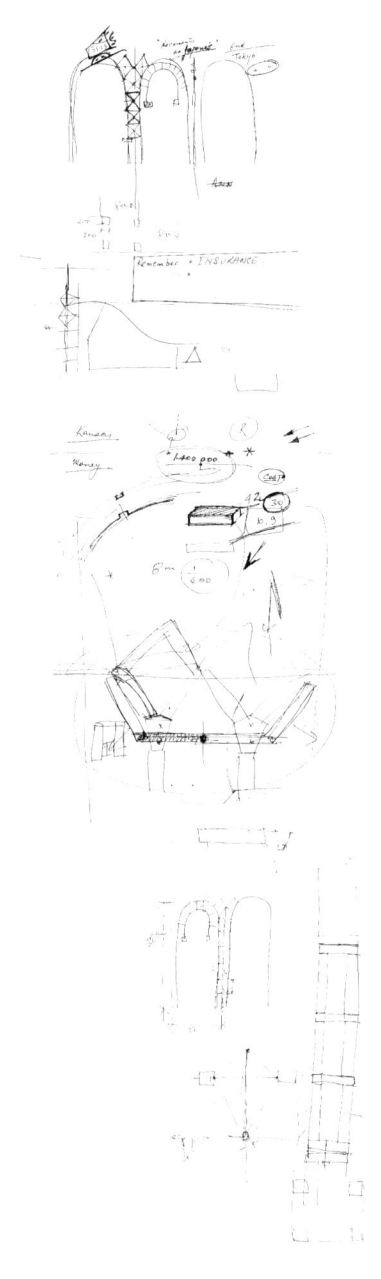

Working sketches for the façade.

builders who built using stone cladding worked to very tight tolerances and very high levels of accuracy. If it were possible to harness the accurate stone-cutting which is necessary to obtain the large stone slabs visible on many post-modern façades, stone elements could then be used as blocks in a pre-cast concrete-type construction. We discussed this with some stone experts who normally worked on façade design and they corroborated our line of thinking. Yes, they said, one could expect to cut stone to tolerances of less than half a millimetre. Furthermore some of the best and most adaptable stone-cutters were international companies which were regularly making elements for complex façades. The component elements of a solution started to appear.

The next task was to choose a stone. The granite of northwestern Spain seemed an obvious choice. It has an interesting colour and is of very consistent quality. We had decided early on that granite was probably the best stone to use because, being igneous, it would offer constant properties which could be exploited to guarantee the quality of the assembly.

The search was then on for a form for the façade. From the beginning we had postulated that to justify the façade it should be used to support the roof of the Pavilion behind. At a philosophical level we theorized that the façade or screen should be like a modern ruin, like a fragment of a viaduct or the aqueduct that we had found in southern Spain. And the notion of the aqueduct gave us the idea about the form. A series of arches seemed a reasonable and logical form for the screen to take.

At around this time we were told that the standard product of the quarries was blocks of stone 20 centimetres by 20 centimetres by 1.4 metres high. We decided to make a prefabricated element out of these blocks. This would be made by epoxy jointing together the pieces, a technology also well known in the stone industry in northwest Spain. These fabricated elements were chosen to be 5 metres long by 80 centimetres by 80 centimetres in overall section and made up of the sub-units. These would then be treated like precast concrete units and assembled into the façade in a similar way as precast concrete elements might be.

The form and shape of the arches were derived from the best shape to support the roof load. By making the roof load tie-support system geometrically similar to the stone arch geometry, and by then attaching this to the stone arches by radial ties, we ensured

Stone:
Pavilion of the Future,
Seville

Peter Rice (back to camera) inspects prototype of stone unit at the stonemason's workshop near Vigo, Galicia, Spain.

Open column unit in Rosa Porina granite.

View of façade looking south-west during construction.

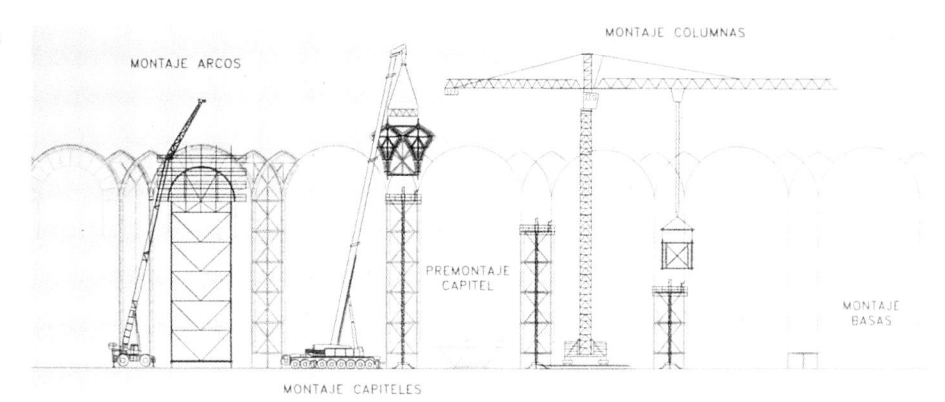

Contractor's drawing showing proposed construction sequence.

Inspecting stone unit at workshop.

Open arch unit after assembly at stone workshop.

that any change in shape of the stone arches was followed by a corresponding change in shape in the support system. This would guarantee that the loading system would always remain funicular, or of the same form and shape as the stone under geometric change caused by wind or other non-symmetrical load parallel to the line of the arches.

The final stage in the design decisions was the assembly and the bracing system. The arch units were assembled as though they were precast concrete units with mortar joints. The pre-assembled units were then treated as single blocks of stone and the granite in the preformed units was analysed and checked to ensure that the 20 centimetre by 20 centimetre sub units were not overstressed during or after the assembly. The steel tie-bracing system for resisting out-of-plane loading (that is, the loading normal or at right angles to the plane of the façade) was designed to treat the stone columns as chord elements in a vertical cantilever, with the shear-resisting elements being the steel bracing system. The form and geometry of this vertical bracing system were chosen to make the least impact on the stone plane itself. By placing the bracing between the vertical stone columns but not in the vertical plane of the façade, we were able to avoid the feeling of a façade flat-braced in the vertical plane.

The last element in the story was the analysis method which was developed for the behaviour of the stone arches. We studied the theory of the way in which stone arch bridges worked, notably following the methods developed by Professor Jacques Heyman of Cambridge University. In his model the stone arch voussoirs, when subjected to tension at the joints, opened up on the tension side, as either the arch extrados or intrados. Normally in stone arch bridges, the large self-weight of the arch is sufficient to guarantee stability under different asymmetrical live loads, usually traffic loads. As our main joints between the pre-assembled sub-elements had no tension capacity either, and the principal loading or pre-load came from the applied load of the roof, the conditions were very similar to the arch bridges treated by Heyman. We therefore developed a computer modelling system which simulated the opening of the joint if it had to carry tension on either its extrados or intrados. This non-linear mechanical behaviour we called flip-flap and it can generate geometrical change within the arches' mechanical connectivity when the thrust line deviates substantially from the arch centre line. The arches were analysed under a full range of loading condi-

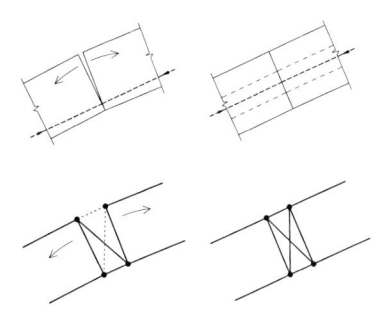

Stone: Pavilion of the Future, Seville

Computer modelling of 'flip-flap' behaviour of joint within stone arch as used for Pavilion of the Future.

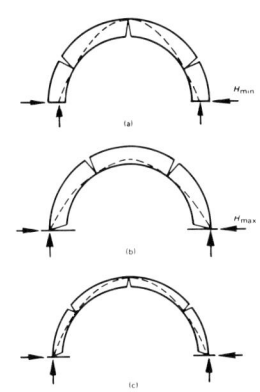

Diagrams from Professor Jacques Heyman's book *The Masonry Arch* showing occurrence of hinges in arch depending on the thrust-line position.

tions, including wind and earthquake, using this programme.

This method was a development of the special analytical system which had already been used extensively for other structures with non-linear behaviour, such as nets and tent structures. This programming system method, known as dynamic relaxation, as originally invented by Alistair Day, one of the team members, is a very powerful tool for examining the behaviour of non-linear, geometrically variable systems.

So, if we go back and examine this innovative façade and its use of stone, we find that at each stage the spur to proceed was created by something that already existed. Each step was prompted by something similar which could be used as a guide. Innovation here was the development of existing ideas and the belief that they were relevant and applicable in the structure we were exploring.

Perhaps the missing ingredient is courage. The courage you need is the courage to start. Once launched, then each step can be evolved naturally. Each step requires careful examination. The courage to start and an unshakeable belief in one's ability to solve the new problems which will arise in the development are essential. It is important to emphasize here that the team should have at least three or four members capable of contributing at every stage of the development. Every stage of the design should be subject to detailed scrutiny by engineers who feel themselves to be sharing in the responsibility. Nothing must be left to chance. Others not so closely involved must also be asked to review the project to question the assumptions and demand explanations. This is obviously easier when one has a large reservoir of skilled and talented engineers, as one finds in Arups. But they can be found even in smaller design groups, such as RFR in Paris. The presence of a competent, dedicated and sceptical checking authority is also very important in this respect.

The final stage in the story is the contractor. When working in a place where one is known, convincing the contractor of the viability of the design is not so difficult. But this was Spain, where I had not built before.

To convince the contractor and the client we made all our calculations available and then we produced a detailed erection method for the arch system. At first the contractor, backed by the client, said that our proposals for erection were far too complicated and would not work – a standard reaction. They proceeded with their

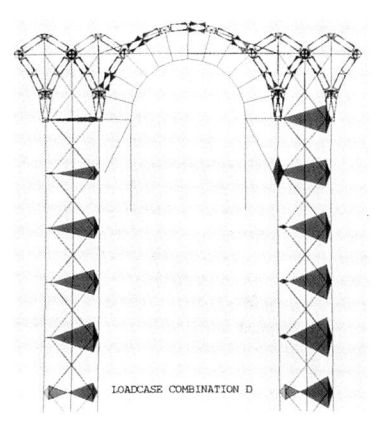

Graphic results of computer analysis of arch under unequal suspended loads producing bending in steel cross-bars of supporting towers.

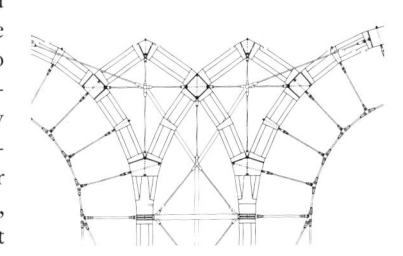

Elevation detail of façade arches.

An Engineer Imagines

Construction study model of the façade showing standard five-metre tower unit assembly.

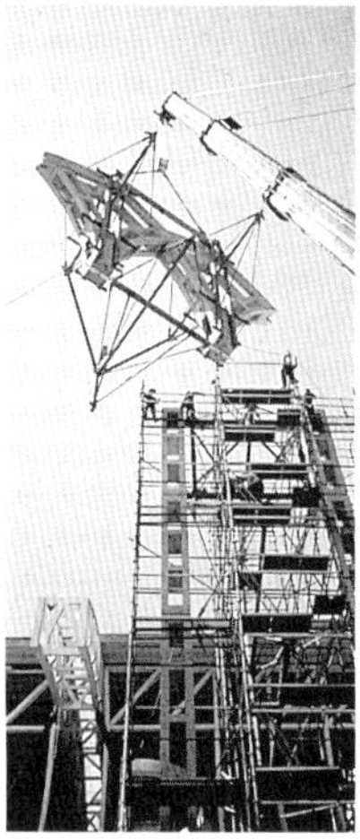

Lifting of capitol unit on top of façade tower.

Detail of joint between stone column units within façade towers showing insertion of bracing steelwork.

View of façade tower with capitol unit placed prior to arch erection.

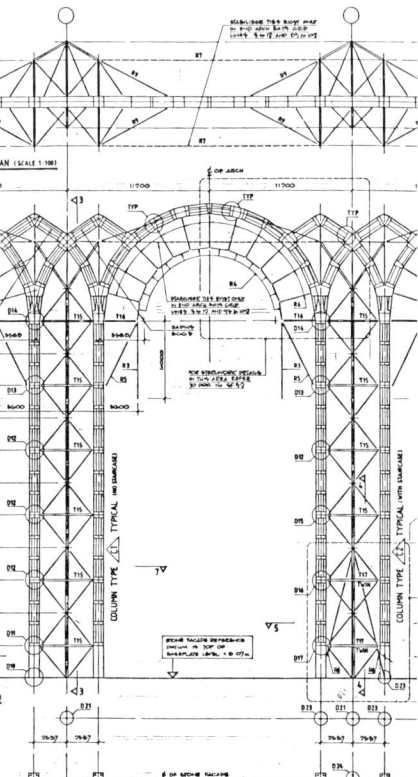

Extracts from working drawing for façade structure.

Stone:
Pavilion of the Future,
Seville

own proposal. This seemed fine to them, until some cracking became apparent in one of the large stone pieces during an early erection stage. We had originally proposed a trial erection where these problems would be ironed out, but, as usually happens, no time was available for this work.

After enormous initial scepticism about the feasibility of the design, the contractor generally relaxes and then makes all the mistakes which he feared in the first place. It is very important that a detailed analysis of what could go wrong exists so that dialogue with the contractor can be continued and that faith in the design is not lost. A cool nerve and an ability to examine the problems logically and carefully are essential at this stage and, once achieved, dialogue is usually much easier, because the contractor's greatest fear is that once things start to go wrong they will be left with all the financial and engineering responsibility of sorting out something beyond their understanding.

This question of ultimate responsibility is a vital one, particularly if the design is unusual and is using materials, such as glass and stone, in an unusual way. We, the designers, must never forget that we do not build our designs ourselves. It would be much easier if we did. Nervi, the great Italian engineer, or Prouvé, were able as contractors to correct any misapprehensions when they encountered them. We have only our power of reason to anticipate and avoid any mistakes, but if an error does occur it must be evident from the very beginning that we will be there to take our share of the responsibility and that we will always be available to see our work correctly completed.

This may not be easy, as the contractual conditions in different countries do not always allow for the presence of the designers throughout the construction process. It may be difficult to persuade the client to pay for one's presence, particularly if the contractor perceives this as unnecessary. But persevere we must because with any other scenario disaster looms. The key thing is that people never lose confidence that the work can be completed correctly. And being there to help and resolve the problems before doubt can take hold is vital.

Needless to say, all these problems did arise on the Seville project, but we were present on site and caught them before they could start to erode people's confidence. In these circumstances it is not going to be all bad news. Some anticipated problems do not arise,

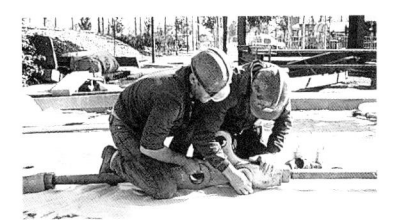

Assembly of arch tie ring components.

Finished assembly showing tie ring bracing arch.

Close-up detail of steel casting passing into column joint.

because the people one is working with become very involved and perform extremely well. On the Seville project the stone-cutting yard which cut and assembled the stone sub-unit elements exceeded all expectations and some anticipated problems with tolerances and jointing proved groundless. There is, however, usually at least one unexpected problem which catches you out. In Seville it came in the natural rift plane of the stone. Granite, because of the way it forms on cooling, tends to have strength in one direction. Having examined the properties of the granite, one tries to get a granite which is as uniform as possible. Various quarries were tested and one was chosen. But when the stone finally arrived to be tested, after most of it had been cut, its strength was not as directionally uniform as we would have wished. What transpired was that working quarries were chosen to give the maximum difference in their directional strengths, because that way the stone is easier to quarry and it does not affect the way it is usually used. Needless to say, no-one thought of this until it was discovered, when it was obvious to everyone. Luckily, at Seville the difference was not such that we could not use the stone. We just had to be a bit more careful and do a bit more testing during assembly, but it is a cautionary tale and demonstrates why one must always be resourceful.

I hope this brief description and explanation of the Seville project demonstrates that there is nothing mysterious in the process of innovation. What is needed is just courage, care and attention to detail, and above all belief and getting started.

The Critic and the Photograph

As an engineer working with architects, I am constantly amazed at the degree to which photographs of architecture dominate what the public – and indeed many architects – see and perceive. The photograph is at best a subtle and effective interpretation of a building by a skilled photographer. The building itself, in all its three-dimensional glory and context, will be something much more subtle, something much more complex than any photograph can or should be able to show. The photograph is a kind of tyranny which often obliges an architect or designer to characterize a building by some specially taken shot, which may produce a memorable image, but which almost certainly undervalues the building and its various plays of light and shade.

Photographs also discriminate against those elements which an engineer can bring to architecture. I also feel that the same kind of simplification can exist in the need to find an epigrammatic appreciation of a piece of architecture. But the real constraint comes from the photograph, which, no matter how good it is, cannot record anything other than what is seen.

For some time I have pondered on the role that the photograph plays in architecture. For an engineer this is particularly significant. Much of what we do, and the special quality that thoughtful engineering can bring to a project, is not photographable. The photograph records an image, a two-dimensional image, one where the three-dimensional nature and the hierarchy of structural or other engineering elements is difficult to convey.

Because the photograph of a building is often the first image and usually the definitive image people have, it is difficult to get people to see those other more variable and subtle factors which light and the texture of materials can bring. The public – even the professional public – will have formed their opinion of the work by studying the images in architectural magazines.

This simplification of the real complexity of a piece of architecture has had a profound effect on the development of current architectural trends. Architects have got to be aware, from an early stage in the development of a design, of how a building will photograph. Unless a firm image can be established which identifies and formulates what the salient features of a building are, it will often get ignored by the public at large. To introduce into this background concepts such as texture and the essence of the material can be very difficult and almost counterproductive.

An Engineer Imagines

Fleetguard factory, Quimper, France.

A photograph requires mass surface to photograph. This was brought home to me when we first built a tension roof, such as the roof of the Fleetguard factory in Quimper, France. When you go to see the factory in real life, your eye picks out a detail such as a node at the end of a compression member and sees the rest of the structure within the perspective of that detail. The difference in visual weight given by the eye and the uniform pattern of nodes and members given by the photograph create different perspectives.

It is quite simply impossible to understand and judge a building, like the Fleetguard factory, by looking at photographs. The true three-dimensional nature of the building is inevitably lost in the photograph. The photograph picks up surfaces, it cannot pick up and register the air between elements and the relative relationship which the pieces have one with another.

Centre Pompidou is another building which has suffered greatly from being impossible to photograph. The building seen from the piazza is a series of planes defined by the structural frame and by the presence of people on the façade. What the photographer does instead is to pick a characteristic detail and elevate it to the image of the building. This leads to people no longer trying to understand the total interrelationship of the pieces, how they are composed and relate one to another.

Because of this, it becomes almost impossible to illustrate a building, particularly a building with a powerful image such as Centre Pompidou, with an element of surprise. That vital characteristic, which can make perusal such a rewarding feature of visiting a building, becomes distorted as one searches for the definitive image.

That other critical characteristic which we have as engineers, and certainly something that lies at what I believe to be at the centre of my own contribution, is materiality. Texture is wholly impossible to register in a photograph. To understand materiality in a piece of architecture, one has to be able to experience the nature of the building itself and the way in which the materials are used.

In times past the role of the photograph was played by the formal drawing of a façade or building element. But, with the formal drawing, one could distort both the colour and perspective of what was being drawn to highlight or make evident a particular effect. Even formal drawing has been diminished by the photograph and is often nowadays reduced to a photograph of the model.

At all levels therefore the photograph has had a detrimental effect on the development of architecture and its appreciation by the public. The photograph is such an ubiquitous element of modern life that many people presented with something, particularly something of a certain size, instinctively compose the photograph in their minds rather than examine its reality.

The photograph is a feature and factor in modern life which is not going to change. And photography, as an art form and as a means of exploring the nature of things, obviously has its own validity. What concerns me is how it has taken over and distorted the actual appreciation of architecture. There is the belief that everything can be photographed, that it is just a question of getting the right angle. The drama of certain solutions which may depend on an appreciation of stability and of the way that stability is achieved are clearly impossible to convey easily by a photograph.

The way a structure is put together or detailed may be used to enhance a particular effect. An example would be the way that the detailing of the Patscenter project in Princeton, New Jersey, by using flat plates for the joint at the top of the A-frame draws attention to the planar non-braced nature of the frame and enhances the perception of the planar quality both in photographs and in reality.

It is true, though, that the architectural concept for the Patscenter project was ordered in a way to make it possible to photograph easily. It is a series of planar structures, repeated at intervals along the length of the building, giving a modular structure which can allow the building to be extended if required.

Another building, the Seville Pavilion of the Future, which used the same structural principle, where the structure is braced through the geometry of the column support system, was not so easy to capture in a photograph. That building, which was lighter and more

An Engineer Imagines

Patscenter factory,
Princeton, USA.

daring, has to be visited to be comprehended and, even when visited, has to be examined from a number of angles before its full impact can be appreciated.

Issues like this, and the way that photographs precondition one to a particular view and arrangement of a building and its elements, mean that the photograph is a tyrant which is very difficult for architects or others working on the design of a building to escape. Indeed I often wonder whether the photograph does not force architecture into a straitjacket, the straitjacket of being photogenic, or at least comprehensible through photographs.

Take, for instance, the post-modern style. Can we say that the strong comprehensible façade images which characterize this style are not influenced by the fact that most people eventually see a building through its photographs?

I have a memory of a building designed by an eminent post-modern architect in Houston where it is impossible to see the full impact of the façade and its composition because of another smaller rectangular building which has been built in front of it. In Houston there is no planning control. All the people who had come to see it manoeuvred until we had an unencumbered view of the majestic composition. The building was thirty storeys tall. This habit, especially prevalent in America, of making a building into a separate entity makes it very difficult for the engineer to contribute much outside the 'make the building stand up' syndrome. The features which matter are purely photogenic and superficial and therefore bear no relationship to the engineering qualities.

I have included the critic in the title to this chapter because I believe the way in which the critics concern themselves principally with a building's image has contributed to this lack of understanding of the engineer's role and contribution which we can properly make. The search for an epigram, something we all do when we write, is part of the simplification which takes place when architecture is discussed in public. I find this way of looking at things much more prevalent in Britain than in France, where I can also read the journals. Maybe this is a reflection of the role language plays in thought and the development of ideas in each country.

In France I find that, in the discussion of a design as it changes, the use of words and language is very important. This can make it easier for an engineer to communicate, because the intention behind a particular direction can be discussed before the precise form that

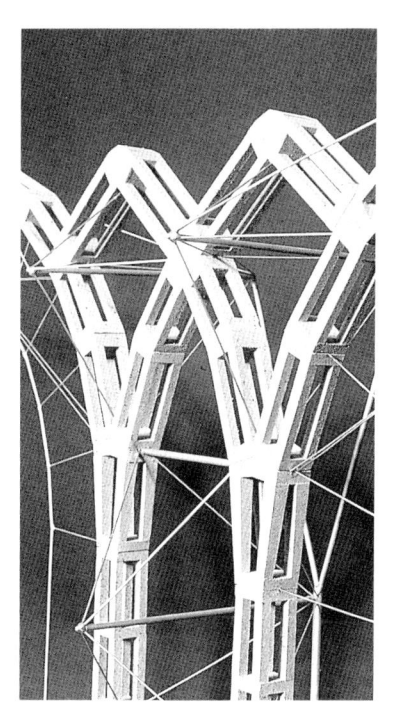

Model of Pavilion of the Future, Seville.

it will take is established. Time and again in France, a design will progress towards a concept of an expressed objective, one where the use of language is the key to identifying what is happening.

In Britain, by contrast, a sketch or a drawing is the usual method by which communication will take place. I find that this can be quite constraining, partly because I don't sketch well, partly also because the sketch can define in much the same way as the photograph the perception of what is happening in too simple a way.

To summarize what is a tricky argument, I feel that the use of photography in popularizing and explaining architecture has made it very difficult for the work of engineers concerned with the use of materials, structure and the character of a building's functions to be understood and explained.

I believe concentration on these issues by those who explain architecture to the public could be very productive, not just because it would help me in what I do but because it would enable people to see the reality of building more clearly.

Working with Industry

Everything we design will eventually be made, assembled and erected by industry. The building industry is one of the biggest and most powerful industries in all Western countries. Indeed modern building is much more a result of developing industrial techniques than a product of designers, architects, inventors or engineers. This is true of all aspects of modern life. To talk of the power of industry is to talk of life in the late twentieth century itself.

As I sit here writing, I am looking out on an English country garden, a haven, one would think, from the industrial monolith. But no. The garden itself is filled with the products of our industrial society. From the obvious ones like lawnmowers we move to the plants themselves, each specially developed; strains for better colour, better taste, higher yield, or earlier planting, abound.

Nothing is natural, an untamed product of the process of natural selection. Even the wild flowers are tamed, classified, and graded so that, when I look for original 'wild flowers' in the hedgerows, I can no longer be sure if any of them have grown naturally or if they have all been planted, or in some way encouraged by enthusiastic amateurs concerned about the beauty of the countryside.

As I look out beyond the garden's edge at the farmland beyond, I see intensively grazed grassland and cultivated fields with barley, wheat and rye in areas which, had it not been for the extensive addition of chemicals, would be low-level grassland for sheep. But the pressure of efficiency and government/European Community policies means that to survive and, especially, to prosper, you must make use of every square metre of land to its most enhanced potential. It is like a race where at each turn a new target is set and the pace accelerates.

I write of this situation with farming because it is a perfect analogy for the building industry. The wild flowers are the craftsmen whose contribution, enriching and variable, is overrun by the steamroller effect of corporate decisions, and who are fast disappearing as a factor in the way buildings are built.

The analogy of agriculture and the building industry can be carried further. Both are large, powerful and political industries, using their muscle to modify the rules to suit them. The most fundamental characteristic to be understood about construction is the political nature of the industry. Permission to build is always a political decision and, in many countries, the large players can have an enormous influence on how these decisions are taken.

The building industry has two quite separate parts, the manufacture of building components and the on-site work itself. The way construction takes place on site is something which is changing rapidly. The high-speed techniques developed and adopted by the American industry give it enormous power to change our environment. The speed of construction means that whole new areas of cities can be rebuilt in a very short time. In defence of this power, it is often said that the Crystal Palace and other Victorian structures were built in a very short time. But at that time large numbers of people were employed on site, so only a few of the large, fast constructions could be built concurrently. Today, with mechanical aids, it is possible to build whole cities in a high-speed way. The industry has the capacity to rebuild all the built environment and, left to itself, it would probably do so.

In Japan, where the industry is politically more powerful than in other Western countries, that is what has happened. Japan has always had a tradition of short-term building, probably because of the vulnerability of older construction to earthquakes. On visiting Japan one is aware of a building industry out of control, where buildings become a fashion item like clothes and individual sites may have a new building every year or so. This incredible capacity is accelerating. The only way to absorb and control this power will be by complexity. Construction will inevitably have to become more complex to absorb our energies and occupy us fully.

Tokyo.

14

The Fiat Experience

In 1978 we were invited by Fiat to look at the design of the car. The opportunity for this project arose because some of the senior members of the Fiat board saw a television programme on flexibility in the home. At the time I was in a partnership called Piano & Rice with Renzo Piano, a partnership which included Shunji Ishida, Noriaki Okabe, Bernard Plattner, Mike Dowd and Reiner Verbizh. This had been formed after our close co-operation on the Centre Pompidou. The intention of the collaboration was to examine building and approach new projects with an engineering bias.

At the time Fiat were struggling with the problem of how to introduce some new thinking into their long-term development programme. It might be asked, what had an architect and an engineer steeped in the traditions of the building industry to do with designing a car? And the answer would have to be 'not much'. But then it was this 'not much' which attracted Fiat management. They thought: 'Who have ideas and who are able to realize them? Who have none of the prejudices of the industry and who could at least look at our problems and tell us what they think?'

Neither Renzo nor I had any great past love of the car. We were not, unlike many of our contemporaries, filled with a passion and opinions about cars. I, for one, had almost no view about cars at all.

We brought to the project an open mind, and a capacity to absorb and examine the information on its merits. This too, we discovered, was an advantage because almost everyone else knew the answer before they studied the problem. What we found was an industry and a way of working which was almost the exact reverse of what we were used to.

In architecture and building almost every problem is new and the solution in one sense or another unique. We were used to composing the problem, to finding out what the most salient points were and then trying to organize the solution to ensure everything was resolved. What, I hear you say, is so unusual about that? What is unusual is that every problem has its unique aspects, which makes an objective study of the facts necessary.

In the car industry it is completely different. To begin with the car has many complex criteria and regulations to satisfy, which are always the same. And it has all its antecedents, both within a company like Fiat and among its competitors. This means that a car is not so much designed as developed. Each new car starts with its predecessor. Indeed the word 'design' for a car is a euphemism for

Peter Rice and Renzo Piano working on a structural model.

Road testing the Fiat Ritmo.

styling and has little to do with the development of a car's performance or engineering. The development of these performance-related elements is a very complex affair, taking many years to bring about a new model.

We were clearly not equipped to design a new car. We were in a way making proposals for a concept car, suggesting ideas that might be incorporated into production at some later date. Even doing this was a departure for a company like Fiat. Normally this is the responsibility of the research department but, even here, there are people steeped in the lore and traditions of the company. They know everything, and their natural and instinctive response to every suggestion is 'But …'

What then could we offer? We started by trying to understand what the characteristics of a good car were considered to be. The most important characteristic we were told is torsional stiffness. 'Ah,' I thought, 'that must be to do with driving performance.' Well, no, we were told that it was more fundamental than that. It is a well-known fact that if a car does not have a sufficient torsional stiffness between the front axle and the rear axle of at least 100,000 kgf.m^2 for a 2.448 metre wheelbase then it will have serious fatigue problems in use.* They then produced a chart and some graphs to show that every car manufacturer achieved at least this magic figure.

This is another aspect of the way the industry works. A company like Fiat is not just steeped in its own traditions. It knows and understands everyone else's as well. The first thing it does when any manufacturer produces a new model is to buy fifty examples, and then take them apart to see if the competitor is doing anything new or unusual, which might change what it is doing. This means that on every proposal there is a detailed history to be compared.

One must never assume that the design or engineering choices made in any car are related to the capacity of the manufacturer to make or properly design any detail. The manufacturer already has a detailed knowledge of the work of everyone else. Not to do something is a matter of choice, a view of what people expect of a Fiat. This image or belief in the marketing image which the public, the customers, have of the essence of the product is a key component in the progressive development of a manufacturer's cars. Fiat is a successful company, so there was no ambiguity about the rightness of what it was producing. But its management could see times were going to change and they wanted to be in the vanguard of any new

Frank Stella: Groningen Museum
Sophisticated computer graphics allowed Stella to participate directly, transposing the free forms of his model into the development of the disciplined geometric forms needed to build it. The project consists of two intersecting leaves floating over the gallery floor. Each leaf has the same undulating geometry, and one is defined from the other by a rotation and a translation.

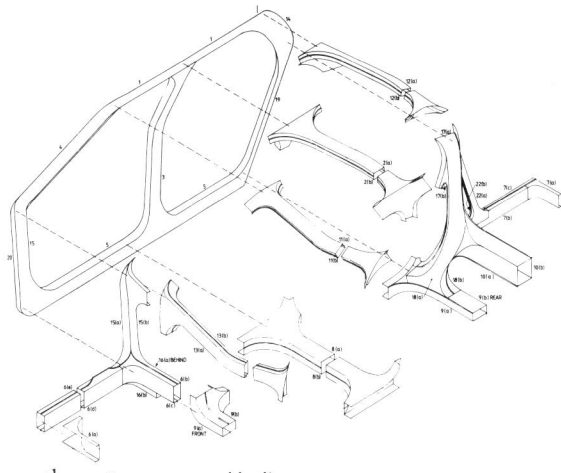

Structure assembly diagram showing how pressings fit together.

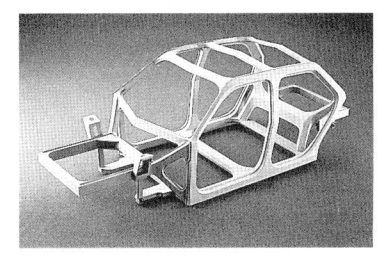

trends which might develop. They wanted clearly to preserve the perception that any decision on new development had been made because they had judged it right, not because they had been forced to adapt to change brought in by others.

The particular stimulus for the need to examine the car at that time was the energy crisis. It was a time when all the talk was of petrol consumption. And there was the growing threat from the major Japanese and American companies to the European and international car markets. Fiat wanted to be state of the art.

When we examined the car and its torsional stiffness we discovered many unusual things. Firstly we discovered that the theoretical stiffness and the actual stiffness were not the same. The theoretical stiffness was much greater, maybe five or six times the actual stiffness measured in test. But then it depended on what was calculated and what was tested. The theoretical value was calculated using the frame or shell of the car without doors or windows. The difference between the theoretical and actual tested values arises from the inefficiencies of the joints. The central vertical door jamb and the area around the windows are highly profiled. When these meet, they make a very inefficient joint but the calculation assumed these joints to be fully efficient, because the calculation of the real joints with plates in many different planes would be very difficult and of limited value. There was much to test. When we talk about the fuel efficiency of a car, we are talking about the weight of the car. At the time, current wisdom was that by replacing the large steel body panels with plastics, one could make a substantial weight saving and hence improve the fuel consumption figures.

Pressings assembled to produce steel frame.

Our proposal was to make a much more efficient base frame, a new chassis, clad in plastic panels. In this scenario the frame would provide the whole of the torsional stiffness. This was possible because we proposed changing the profiles so that the door closing section was separate from the structural section. The base frame was clad throughout so that the structure and the car body had the same relationship as they might have had on a building.

Nobody is going to pass their power base to others. Within the company the transfer of information between departments was very sparse. I think that this is characteristic of all large companies, not just Fiat. The development and transfer of information is done defensively. People are given the minimum possible details and this makes new ideas difficult to introduce. Unless you can introduce a

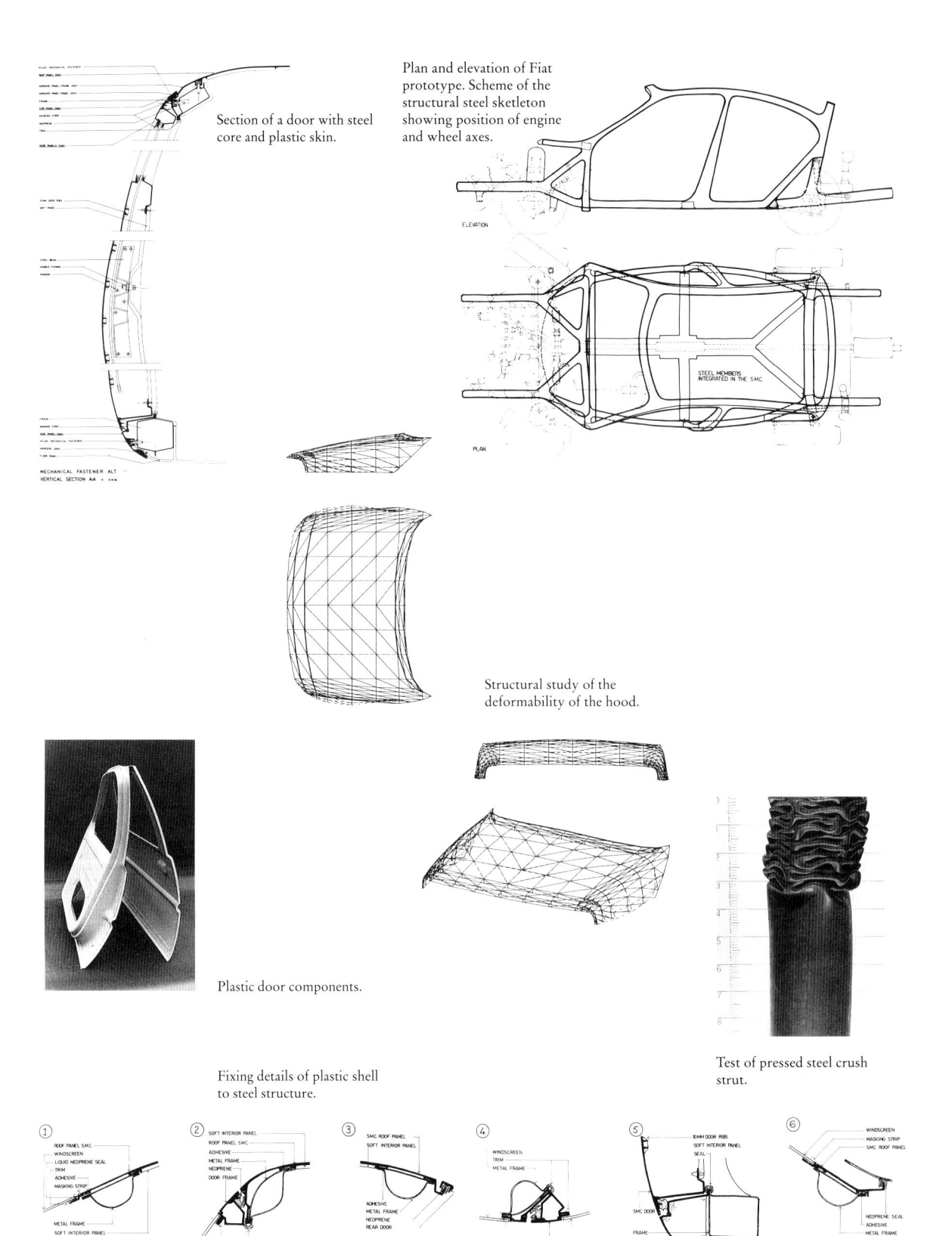

Section of a door with steel core and plastic skin.

Plan and elevation of Fiat prototype. Scheme of the structural steel sketleton showing position of engine and wheel axes.

Structural study of the deformability of the hood.

Plastic door components.

Test of pressed steel crush strut.

Fixing details of plastic shell to steel structure.

new idea with the confidence that it will work, it is difficult to make progress. We were exempt from this constraint because, on the simplified models we were using, we could do all the calculations and proof ourselves. The technology itself was not exceptional. Standard computer software existed for most of the tasks we wanted to do, so we were able to make the proposal and show that, in its fundamental form, it satisfied safety and other structural criteria.

Initially, when Fiat had proposed the project we were given six months to come up with ideas. After the six months had elapsed, the Fiat people judged that the proposals we had made were of sufficient interest to warrant further study. To carry out this work, they set up a separate research institute owned by Fiat, called IDEA (Institute of Development in Automotive Engineering), under the general control of Francesco Mantagazza, a peripheral Fiat man, known to members of the board. Renzo was chairman and I was vice-chairman.

IDEA was housed in a villa on a hill north of the Po in eastern Turin. This villa became our working home, the most sublime working environment I have ever experienced. Out in the park, which surrounded the villa, were orchards and vine walks and a large area liberally planted with fig trees. I never had difficulty tearing myself away from London.

Fiat seconded to IDEA a group of specialists who could advise us on aspects of the design with which we were not familiar. We too collected around us from universities and polytechnics another group of people to help us with elements like wind drag, acoustics and so on. They say a little knowledge is a dangerous thing. Armed with this support system, the specialists provided by Fiat and IDEA were mostly young, hardworking and dedicated professionals making their way up the profession.

Our proposal was not confined to input on the structure and the body. One of the principal problems of the car is the various functions which must be fulfilled by the paint scheme. It provides the outer visible finish and the corrosion resistance. The two are largely incompatible. The quality of paint needed to make a good visual appearance is not good for corrosion resistance. Also the whole car must be painted at one time, because even very minor differences show in colour or finish between parts of the car. Indeed there was an alarming failure rate on visual inspection of cars which had been painted, apparently an industry-wide problem. Our proposal involved separate painting of the frame and the body which no longer needed corrosion protection.

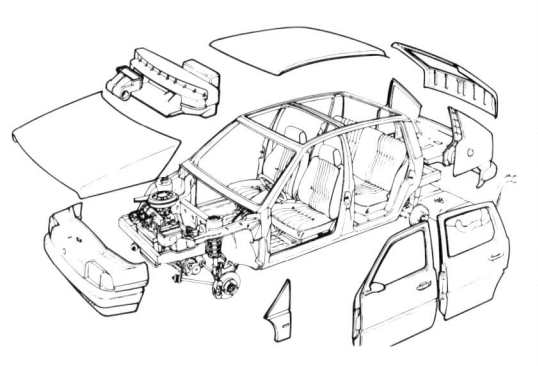

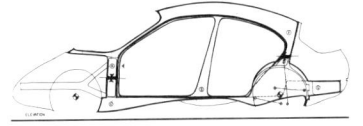

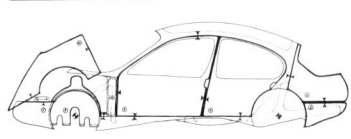

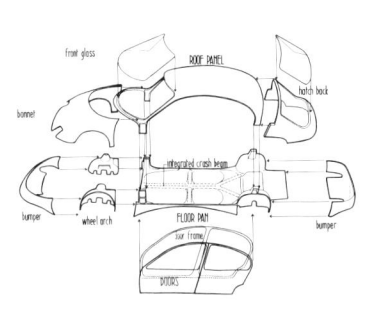

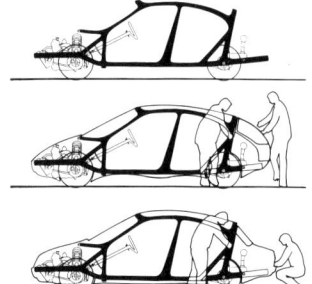

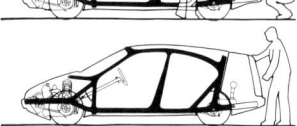

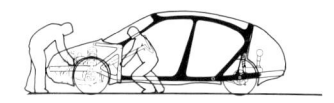

There were other aspects to the proposal which were generally considered to be interesting. All the new ideas were in the frame and in the body. Some of these were carried forward into the production process where some serious anomalies existed as well. One of the most surprising things to amateurs like ourselves was that the front seat can cost as much or even more than the engine. This is because the front seat has a large craft element in its manufacture and must be fitted after the car has been painted, that is after the front doors have been hung. This adds considerably to the cost, as fitting in this confined space is expensive. In our proposal this might not be necessary (we were not sure), as the panels could be painted together before assembly.

The combined proposal was presented to Fiat after one year of the study. Some of the apparently awkward details, such as the connection between the frame and the body, had side benefits. This made the car much quieter. Normally the internal noise in a car is caused by panel vibrations, triggered by the direct vibration connection between the engine and the body panels, by tyre vibration and via the springs. All our separated connections meant that the direct vibration links did not exist in the car when tested in prototype. Fiat, especially the engineers, called it a house on wheels and that, in a way, was what was intended. While I had been working on the structural and body concept, Renzo had been working on the use of the car as a flexible living component. He proposed that a variety of different body shapes could be used with the base frame, giving the car-owner the option of changing the shape and internal arrangement of the car for different occasions. The basic strong, highly resistant frame was, like the engine, a permanent part of the car and the rest was part of a flexible lifestyle approach.

Between these interesting, albeit naive, ideas and a real car, there were many steps to be taken. But Fiat was interested and decided to continue on the project. At that stage they decided that it had better become a Fiat-only project and we were no longer required.

It was sad to finish but it was clear that in the real development, even of a prototype, intervention by us would have been very difficult. The difficulty that we would have had continuing on the project was illustrated for us shortly before we finished.

We had a team of ten or twelve people working on the project at

Studies of the use of the car as a flexible living component. A base frame to which the owner can attach different body shapes to suit individual lifestyle requirements.

IDEA with different people responsible in broad terms for sections of the project. Fiat had decided that, to explore the use of plastics, it would work with a major manufacturer, General Electric of Holland. The Fiat management told us that they would call a meeting with General Electric to start exploring the use of special plastic components. The meeting was arranged for ten o'clock on a Tuesday morning. The GE people turned up. Renzo and I entered the meeting, believing we would be running it, as indeed we were if you looked at the seating. Suddenly we realized that we were in the middle of a large discussion where everyone had his counterpart. The bumper man from GE was sitting opposite the bumper man from Fiat. They knew each other already. They had a lot to talk about. We were the outsiders. There were seventeen representatives from each company in matched pairs just like tennis. As they knocked up, we watched with a detached awareness that our role would be difficult if not impossible to sustain. A group of ten people cannot interface with a group of 100 who have separate specialists for every aspect of the design: at the early information-gathering phase of the project, yes perhaps, but at the implementation stage it was clearly going to be impossible. It was clear that our days were numbered. We had made our contribution. The project for the prototype was launched. Our role was complete.

Fiat did in fact proceed with the project and made the prototype, which gave enough encouragement for some of the ideas to be studied further. Indeed, some of the ideas proposed have started to appear in production.

The most important thing Renzo and I learned during this time was how Fiat itself and other similar large companies worked. In large companies, the company's structure and an individual's place in it are more important than the opposition. Every product that the company makes takes a long time to develop. Comparison is made with the decisions adopted by all the rivals. The essence of preserving the character of the company in the product is essentially a decision made by all the participants, in the presence of and with the collaboration of members of the board. Responsibility is shared. In the technical departments, knowledge and expertise comprise one's power base and one's power base is something one guards very jealously.

Every decision is scrutinized very carefully for its defensibility. Big companies with their large capital bases are considered fair game by lawyers and the public at large. Their defensiveness is something we all benefit from generally. It has produced the reliability and safety of cars as we know it today. However, the whole atmosphere inhibits the free flow of information.

At one stage we asked whether the torsional stiffness should change between a front-wheel-drive car and a rear-wheel-drive car. We were anxious to know because we were proposing a concept where the fatigue criterion might be less significant. Was there a case for modifying the stiffness? There was no answer. Clearly neither Fiat nor any other car manufacturer had ever made cars with a lower stiffness. To do so would be to take an enormous and unnecessary risk in the battle of liability.

This defensive approach means change is difficult, if not impossible, particularly on those items not directly related to marketing and styling. The changes that we in the Western countries (including Japan) must make in the role and performance of industry in our society will be very difficult to bring about. Industry is responsible only to itself. Change can be brought about either through regulation or by a change in the public's perception of values. Regulation either at the national or supranational level is doomed to failure. It will only reinforce the company structure by requiring more expertise and by further boosting company structure which sees change as a slow development and quiet addition to the existing product range.

What is needed is the inspired invention introduced by Alex Issigonis in the Mini. The engineers have got to be allowed to think freely and inventively. We have to find a way to introduce more individuality and craft speciality to make the car a product where there is some reminder that it was designed and made by people for our collective benefit. What that reminder should be may change in its details.

As we survey a world racked by pollution and by our capacity to build, and as we fill it with products to the point where there is no longer room for the original inhabitants, education and awareness are the key. We cannot expect industry to change of itself, because the legal framework gives no incentive to initiate change.

Another important feature of industry is its rolling investment programme. At any given time there will be large investments in land and production methods which must be amortized. No time is a good time to start. There will always be something in progress which must be finished. This, coupled with the complex interaction between management and labour, will always make changes difficult to introduce.

But change will come. Computers and the way that they can assist management to handle much more complex situations will make that inevitable. It is important that this possibility for change is not wasted on goals which will lead only to short-term gains and not to the real benefits we so desperately need. If society can properly identify the goals and engineers are given their head to invent and innovate, then I believe all is not lost. But the change in thinking that is needed is great. As Ove Arup said in his last address, engineers must accept that responsibility for change lies also with them. They must use their muscle as well as their inventiveness to encourage the kind of change that only they can bring about.

Final prototype Fiat.

View of the Grandes Serres at the Cité des Sciences et de l'Industrie, La Villette, Paris.

The Chameleon Factor

I have often wondered how I can find myself working equally happily with a variety of very different architects. Indeed I often find I can be working on the same day with two or three architects who would find each other's architecture very difficult to accept.

At first, when I started thinking about this problem, I was puzzled myself as to how the co-operation between architect and engineer works. We have said already that the engineer is working with the physical performance of materials, with structure, or light, or even the acoustic performance of a space. These are measurable physical features of the architecture.

I find that I actively enjoy understanding the aesthetic position of different architects and exploring what that means in terms of how one should proceed with a solution.

There is a certain romance in an architect's philosophy. Architects usually feel very strongly about the importance of certain features in the way they develop their projects. Nowadays I am usually involved in projects because the architect feels that I and my team can improve, or in some way enhance, the architecture as a whole.

Strong architects usually have a clear view of how they would like that intervention to be seen. This is best explained by example. Two examples which are very different are to be found at La Villette, the *serres* or glass boxes in front of the building designed with Adrien Fainsilber as the architect and the *galeries* or bridges in the park designed with Bernard Tschumi.

With Fainsilber the architectural intention was to explore the use of glass in a refined and precise way which demonstrated the physical properties of glass, including its strength and transparency. As part of the solution we designed a stainless steel frame and a series of prestressed stainless steel cable trusses to restrain the glass against wind load. These cable mullions have, by virtue of their geometry, two opposing centres of rotation. As a body cannot rotate about two different centres at the same time, the geometric device provides relational stability while appearing to be unstable. However, when working with Fainsilber, it was not the apparent instability of the system which was the important factor, rather the lightness and transparency of the whole façade. The detailing of the cable truss and of the restraining ties for the main frame of the serres was devised to minimize visual impact in order to emphasize the lightness and consistency of the whole. Even though there is the impression of a 'no hands' solution, particularly with the cable mullions,

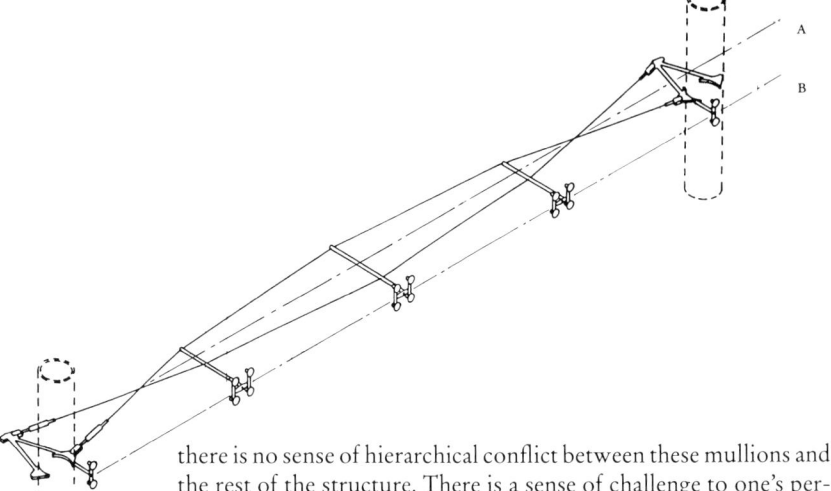

Diagram of cable mullions for the Grandes Serres demonstrating stability as a result of opposing centres of rotation. The combination of these two axes of rotation means that the glass is used to stabilize the truss.
A Axis of rotation of cable truss.
B Axis of rotation of strut at glass plane.

View looking through the Grandes Serres conveying lightness and transparency of façades.

there is no sense of hierarchical conflict between these mullions and the rest of the structure. There is a sense of challenge to one's perception of stability and the structure demands understanding.

A critical element of everything that I try to do is to seek some element in the physical interpretation and presence of the structural form which challenges those who see it so that they come back to see what is really happening. This is what I would call examining the nature of things. It is a study of the nature of the structure, rather than the image, which yields the greatest puzzle and the greatest satisfaction when it is understood.

The Fainsilber serres are in direct contrast to the *galerie* structures designed with Bernard Tschumi. Tschumi seeks to test the viewer in a more direct way. It is the form itself and its juxtapositions which become the challenge. Unlike the serres, nothing is smooth and well ordered. Rather the geometry and the details are set up to clash and provide opportunities for non-standard intersections where members meet. Stability is an issue here too but it is the overall stability of the assembly that counts. Elements are cantilevered to create the maximum sense of distortion and asymmetry. A sensation of instability comes from the feeling that the whole thing may be about to topple over, and there is a slight sense of unease as you walk under the canopy structure. Furthermore the members are arranged to provide geometrical conflict which must be resolved. This is done in an evident way so that there can be no doubt that the conflicts were intentional. An example of this was when the holding-down tie for the *galerie* beam tried to pass through the main support column position. It was necessary to provide a detail which enabled the tie to circumvent the column.

These two examples of different approaches to steel detailing and to one's perception of stability I found challenging and interesting to explore. The fact that they represented two opposing views of architecture added to rather than detracted from the engineering interest. I was fascinated by the juxtaposition of the two steel structures on the same site. Both demanded inventiveness and even innovation. For the engineer the interest lay in the way each solution separately provoked an attitude towards the way one handles steel.

I find the approach of Bernard Tschumi interesting because it questions all of the normal conventions one uses when designing steel structures and in a very deliberate way. One provokes conflicts, searches for the most unstable solution and emphasizes the asymmetrical aspects rather than the symmetry which is characteristic of the more normal and conventional approach.

I would like to remark here how interesting this is for a structural engineer like myself because, although most structures, particularly those used for architectural effect, are symmetrical, the essential environmental loads – such as earthquake loads, wind loads, snow loads and other so-called live loads – are essentially asymmetrical. Whether this is a product of the time when building materials were very heavy and therefore swamped the effect of all other loads, or because architects naturally prefer symmetrical forms, is a moot point. Suffice to say that it is exciting and interesting to explore a structure where asymmetry is part of the aim.

These two examples neatly illustrate how, to an engineer, the design challenge can be independent of the preferences of the architect and an effective contribution can be found in either case. Indeed in the particular situation just described, I enjoyed the opportunity to challenge the assumptions of the serre designs.

What happens when one works with different architects is that each architect's attitude and opinions become part of the problem being solved. And provided one can introduce – as one almost invariably can – an element of engineering exploration either in the nature of materials, or structure, or light, or some other physical phenomenon, no loss of identity or independence need be implied for the engineer.

I have, of course, my own views of how I would proceed if I were working on my own and these are explored in some of the projects mentioned in other parts of this book. But even here my essential objective is to explore materials, structure or another physical aspect. Therefore both the serres and the galerie structure are projects which fit very well within my general engineering philosophy, even though they appear to be very different.

Galerie at La Villette, Paris, demonstrating visual instability.

Galerie at La Villette, Paris, showing detail which circumvents column.

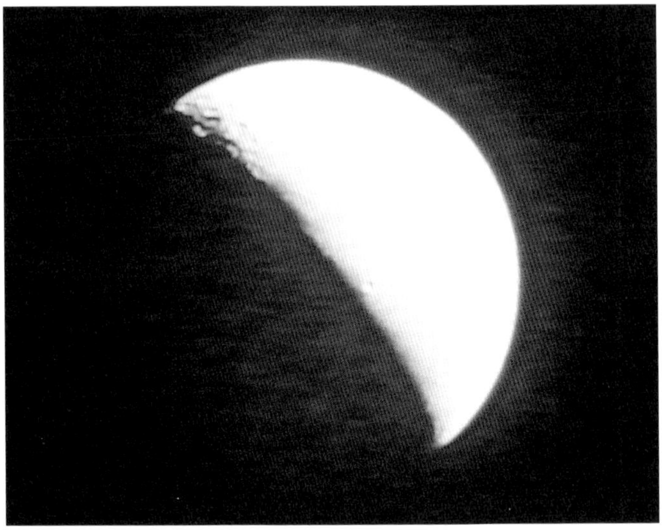

The Full-Moon Theatre

The Full-Moon Theatre is a project about minimum intervention. Myth, fantasy and reality combine under the powerful ritual presence of the moon itself.

The theatre is lit entirely by reflected moonlight. The calculation to track the moon, define the geometry of the reflections which perform different theatrical functions – spots, sidelights, footlights and so on – was developed in London and Paris, working with Humbert Camerlo, a theatre director. Otherwise all the development is the product of craftsmen working on site. Young engineers don't sit and draw in the office. They are all sent to the theatre to participate in its development, both physically and mentally.

A special lightweight mirror has been developed using 1.5 millimetre chemically toughened glass as the base artefact of the assembly. It is and will continue to be a place of experiment in theatre, the psychology of light and the myth of the moon. The committee of the Ateliers de Gourgoubès includes physicists, artists, musicans, writers, choreographers and dancers, film directors and senior television producers, and myself, an engineer, to create a multi-varied framework for research. All original work must take place on site. It is a project at its beginning.

The Full-Moon Theatre is set in the domain of Gourgoubès, north of Montpellier, and is part of the Ateliers de Gourgoubès, a musical and theatrical workshop directed and led by Humbert Camerlo. Humbert has defined the intention and objectives of the Centre:

> The Centre at Gourgoubès is based on a cross-disciplinary philosophy. The projects, workshops, experiments and new pieces are being developed there by artists as well as by scientists of various disciplines collaborating on common objectives. The fields of scientific or technological research and art creation are explored where the various patterns of art and sciences can meet, complement, confront or grow in parallel in a sympathetic environment. To operate in such a multi-disciplinary network has long been a demand from artists, scientists and philosophers and the Centre is one response to those needs. The unpolluted natural environment of the Centre at Gourgoubès is one of the key elements for this kind of integration to be achieved with success and harmony, away from regular city tension and agitation. Each project or activity pursued at the Centre has designed its own way of development and progression.

An Engineer Imagines

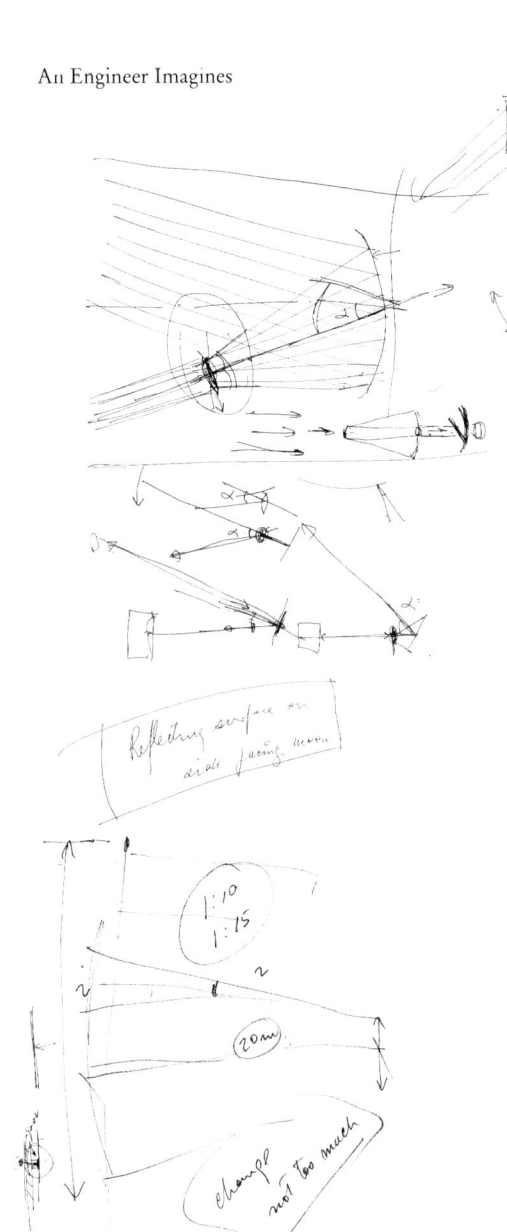

Sketches showing principle of reflecting moonlight on parabolic mirrored surfaces.

One of the most original pieces of equipment that is being constructed today at the Centre de Gourgoubès is the *Théâtre de la Pleine Lune* (the Full-Moon Theatre), the first of its kind in the world today. It is the basic element of the foundation of the Centre and its system of operation, as well as a symbol of integration between art, science and environment.

My role at the Centre is to develop and make possible the Full-Moon Theatre: a natural amphitheatre, entirely lit by moonlight. The light of the moon, even at full moon, is very low, and requires to be amplified many times to create the theatrical experience.

Given these general conditions, the role of the engineer is the reverse of normal. The role here is to enable the members of the Centre at Gourgoubès to make and develop themselves all the different elements of equipment they will need for the theatre to function. The project must remain natural, in tune with the place and its spiritual vibrations. Anything too technological would disturb this balance. Some technical intervention is necessary – mirror technology is one example – but this must be the lightest possible.

What I feel is essential in the Full-Moon Theatre, and in the Gourgoubès project generally, is that everything that happens there is rooted in the place and comes from the hands of people who live and work there.

It is a natural development. It is not something about benefiting from the situation we're in today, because there's no doubt in my mind that somebody could have built the Full-Moon Theatre 300 years ago, as soon as they had mirrors. What amazes me is that inventors and classical astronomers in the seventeenth, eighteenth and early nineteenth centuries, when they were very interested in natural phenomena (witness the Lavoisier and Buffon experiments), did not attempt to do something of this kind, given that the moon has always had enormous mythic significance.

The laws of light and the laws of the physics of light have been known since the time of ancient Greece. Maybe not in the way we know them today, nor in the way they were defined in the nineteenth century, but to the extent that people would have known the effects that can be created with, for instance, a mirror-like object held up in the sun's rays. (During the siege of Troy, Archimedes set fire to the Greek fleet using copper shields.) People have forever been attempting to develop methods, techniques and rituals to

The Domain of Gourgoubès

The Full-Moon Theatre

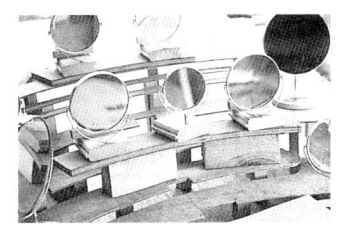

Humbert Camerlo originally tested his concept for capturing and focusing moonlight using concave shaving mirrors in September 1987.

enhance the power of the sun. The moon fulfils just the same role but in reverse. The sun is the energy and the moon is the myth.

The core of the project is understanding how to enhance the natural development; it is actually not about someone like me thinking about the problem, going down, getting some industry there to make everything, organizing construction and saying: 'All right Mr Director, here's your theatre, do what you want.' Here there must be a natural outflow from the place and the people who are there.

It can, and probably will, spawn other developments which may be much more mechanical in the way in which they evolve. But the essential first theatre is about the capacity *artisanale* – craft-based skill – being used inventively by the people on the ground.

If we chart the development of the moon reflectors in the first season they were used it will explain the philosophy and its evolution. The principal reflectors developed were the 'Keplers' and 'Archimedes' designed by my team in London and Paris, and the 'Copernics', designed by Humbert Camerlo. The Keplers were large parabolic reflectors designed to focus the moonlight on an area of about one metre by two metres. The geometry was defined using a timber grid-shell, a lattice frame on a half-metre spacing whose shape was controlled by the length of each lattice member. This was then covered in thin plywood. The plywood, three millimetres thick, was cut to patterns provided from London. Once made, the plywood surface was covered in reflective mirror plastic to provide

Copernic reflectors under construction in the Full-Moon Theatre, summer 1991.

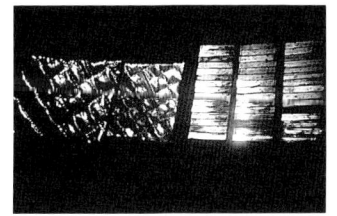

1991: view of the Full-Moon Theatre with Kepler and Copernic prototype reflectors positioned at different levels to avoid shadowing and maximize reflective surface area.

An Engineer Imagines

1991: view from the seating tiers out over the stage and valley to the cleft in the hills, due south, over which the moon travels.

1991: on-site construction of an Archimedes prototype reflector using a radial framework to form the paraboloid.

1991: The framework was then covered with a plywood surface and then a final surface of mirror foil. This reflector was designed for use on the stage.

1991: a Copernic prototype reflector under construction. A steel framework supports numerous flat mirror-glass slats which can be rotated individually while the whole structure can also turn to follow the moon and stage action.

1991: on-site construction of a Kepler prototype reflector using a pinewood grid-shell to form the parabolic curvature. This was covered with first a plywood layer and then a paper layer to smooth out surface imperfections, with a final surface of mirror foil.

1991: Hung in a wooden frame it is able to pivot horizontally and vertically to track the moon and action on stage. The Kepler was designed for use from within the Theatre.

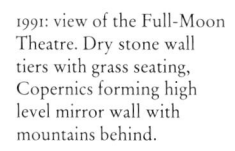

1991: view of the Full-Moon Theatre. Dry stone wall tiers with grass seating, Copernics forming high level mirror wall with mountains behind.

1991: View of the Copernics forming a moonlight-capturing wall on the highest tier of the theatre facing the stage.

The Full-Moon Theatre

FAR LEFT 1991: Copernics multiple angle reflections.
LEFT 1991: Copernics individually labelled control strings turn each mirror slat.

1991: Copernic, Kepler and Archimedes reflectors turned over to avoid focusing sunlight and causing a forest fire in the dry brush around the theatre.

a complete reflector. Initially the carpenters who made the reflectors had difficulty achieving the accuracy needed but after two or three attempts large reflectors of reasonable quality were achieved.

This gradual improvement was the essence of the enabling process where the real expertise on how to make the reflectors was in the hands of the people in Gourgoubès. The principal disadvantage of these techniques was that the reflectors were large, very light and therefore very susceptible to wind buffeting. This wind buffeting was something we had not anticipated. Normally there is no wind at Gourgoubès, but at dusk sudden squalls would blow up and last for maybe thirty seconds then die down quite unpredictably. It's a very hilly place, mountainous, and because it is in a valley the air currents cool off very quickly and differentially; you can suddenly get strong gusts of wind coming up. And all these enormous reflectors blew over just like sails. Humbert had underestimated the impact of the wind there. We were almost at the beginning of a performance and the wind came up and was gone in ten minutes. We had worked it out, put some more reflectors up for the next night and the bloody wind came from the other direction. It was all generated by very fine local factors. That is a problem we have to solve, and we have solved it a bit by putting in bases and giving them certain base strengths. In a sense this detracts from the flexibility, it means that it's awkward – and it will always be awkward – which actually in the end I like. Even if we did try out our technological wizardry, it wouldn't defeat the fact that the place is the place.

And the only people who really understand the place are not people who have visited, but the people who live there. The kind of people who are there participating do not have an instinct for that kind of understanding either: they do not naturally go out and feel the wind. They are having to train themselves to understand what is happening.

An Engineer Imagines

Detail of dry stone wall seating tiers and steps.

Morrocan mason constructing the dry stone wall amphitheatre.

Full-Moon Theatre performance by the American dance company MOMIX, daylight rehearsal.

Night dancer illuminated by a reflected spotlight, necessary when clouds obscure the moon.

Other aspects of the Centre were developed entirely on site. The theatre itself was built by a group of six masons from the Atlas Mountains of Morocco, who came to the site and in eight weeks, under Humbert's direction, built the complete theatre. Local carpenters, masons and metalworkers developed all the artefacts necessary from geometrical information provided by my team in London. In general, the members of the Centre, particularly those who live there, understood this approach. The same was not true for visitors, especially the professionals from television and the theatre. For them, the idea of imperfectly made reflectors was difficult to understand, because they were used to using highly developed industrial products with a predictable performance.

Another dilemma in the middle of all this is that we invariably think too big today. When we came to work on the stage and the seating, the size of the constructed stage was in my opinion too large. The size was defined by what was needed by a dance troupe but we could not light all this area. The moon reflectors are large because they have to collect all the light and then focus it. So, in order to light one square metre to five times incident moonlight intensity, you probably need 6.5 square metres of reflector. Six and a half square metres of reflector are not nothing when, because they are not flat, you have to adjust all the elements with little mechanisms in order to focus them properly.

This was well illustrated during the television presentation, *La Nuit des Etoiles Filantes* (The Night of the Shooting Stars), broadcast from the Centre in August 1991. During the two full moons prior to the August presentation, the television crews came to Gourgoubès to film performances at the theatre. (One was by Momix, the American dance group, whose director, Moses Pendleton, is a member of the committee of the Centre.) The difficulty of getting, by tracking the moon with reflectors, a consistent light on to the stage, made them feel very frustrated. Eventually sufficient consistent light was possible for adequate filming to take place; the technicians and the others who were there had some difficulty understanding the role they were expected to play, acting as labourers to adjust and focus the reflectors. The work was hard and needed constant attention. Once everyone entered into the spirit of the exercise, a rewarding time was had by all.

This aspect of the theatre at Gourgoubès is the attempt to keep its development natural, local and non-technological. Within this

The Full-Moon Theatre

The House and Atelier de Gourgoubès.

context there are three separate research objectives. The first is the spiritual or ritual needs of the Centre. The second is the understanding of the way people react to moonlight, both in terms of its mythical and its simple physical reality. Thirdly, there is the development of the theatrical experience itself.

The spiritual side of the experience is enhanced by the place itself. Gourgoubès lies on one of the famous medieval pilgrim ways to St Jacques de Compostelle, and there is a spring just below the Full-Moon Theatre where the pilgrims used to stop for refreshment and spiritual sustenance. This connection means that one is permanently aware of an extra presence when one is at the Centre.

The house which is now the Centre, at least the original part, dates from about the ninth century. It used to be a farm raising sheep, olives, fruit and cereals. In the seventeenth century it also farmed silk. The basements were laid out for the silkworms. The vegetation was mostly mulberry. Then, I assume, the people went to the East in a big way. At the turn of the eighteenth century, the silk trade gradually went into decline, although it does still exist in parts of France.

The house is big, but it has always really subsisted on agriculture. The trees now include a mix of fir and pine. The most common trees are a sort of evergreen oak, most of which are ancient. Everything is very old, very dry. And trees grow slowly, so a respect for the vegetation is built into the place, because if you cut something down you may have to wait 300 years for an adequate replacement. And that was another problem for the Full-Moon Theatre. We had to have direct passage for the light, so the first thing seemed to be to get rid of the scruffy little trees. But, obviously, even cutting down a small tree had to be thought about.

I went down to Gourgoubès in early summer 1991 when they were preparing the theatre for a performance to be televised. The people there were realizing the difficulties of building accurate optical devices from timber, plywood and mirrored plastic. The reflectors had all been made on site but they were not quite right, and the people who were making them were flailing around in terrible trouble. The people involved – a group of about twenty, each of them coming in a way from a different discipline, included a film-maker, a composer, a ritualist, a manager from Antenne 2 (a French television company), as well as those hangers-on who normally do not physically do anything – had to push and shove and get the thing

Looking from the house, the edge of the Full-Moon Theatre is just visible through the trees.

The huge theoretical paraboloid from which all Keplers are cut. Its axis lies through stage and moon, with the stage at its focus.

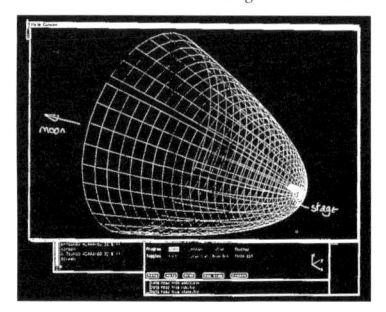

View of Full-Moon Theatre from the moon to ensure there is no overlap of mirrors. A similar check was made from the stage.

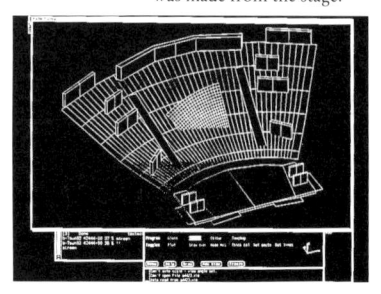

Checking the focus of a Kepler reflector design.

to work because cameras were rolling. And one of them said to me that he could not understand why the reflectors didn't work, because I've got a good reputation for knowing what I'm doing. I had enormous difficulty in convincing him of the underlying philosophy. The carpenters and other craftsmen, they understood because that was how they worked anyway. It was the others, the successful career people, who had enormous difficulty understanding the fact that here we were making these large reflectors and they were not perfect. I tried to explain the strategy to them. Because they were French and intellectual, you could see that they could hear the words, and that they could understand that there was actually an argument there, but then they would look down at their feet and say, 'Comment?' and that brief moment of understanding would disappear and they would be unhappy again. And in a sense that's my contribution to the project, to understand.

Of course, we had to use mathematical models to track the moon, to ensure that it would focus properly, to check the light and to assess the level of light needed. So there was a lot of background calculation to achieve the particular effects that we wanted, and obviously I had Arup people – in particular, Andy Sedgwick – working with me. But that was then put together in the form of intellectual information, which meant that the whole of the physical development could take place in the hands of the people there and could become their expertise, not a product of industry.

Apart from a good clear philosophical position, I think the making of the Full-Moon Theatre is something which energizes people. People become almost mad in the intensity of their commitment to get the whole thing to work. But once you have this I think the place enables you to touch fantasy just because the place is fantastic.

Another very fundamental thing about the Full-Moon Theatre is that it is the reverse of the normal theatrical experience. In a normal theatre you enter a black box, you suspend belief and you prepare yourself for some new visual assault, whereas in the Full-Moon Theatre, the process of getting there, the process of being there is actually what the experience is. You're having to play with the normal theatrical parameters in reverse and it becomes like a pageant. But, of course, that is not what we actually consider as a theatrical event: using the theatre as a place to enhance our understanding of communication. I think this inversion of the normal experience, in which you know where the ambient light is coming from, means that everything is part of the performance. The process of providing the theatre lighting for the performances, rather than being completely hidden behind the mystifying screen as it is in the theatre – 'How did you do that? My goodness!' – is the reverse. You actually have to be involved in how you did that, it's part of the experience too.

The other great interest of the project is people's reaction to moonlight. Many people feel that they are very influenced by the phases of the moon and the full moon is a particularly special period for them. This quite obviously affects how people react to things which are organized to take place in the light of a full moon. In addition, because moonlight is something which is lighting things which you know, it's very probable that you would have already seen these things in daylight; you are actually seeing things that, if the same amount of light was shone on them in a black box, you wouldn't see at all.

So there is this interaction between people and moonlight, and the way in which they react to what they see in moonlight. And if you superimpose that in a theatrical experience when what you're actually shown is unexpected, you have a very interesting and exciting mix. It is not just seeing that the trees are green because you know the trees are green. It's almost a mixture of the scientific and the psychological, described in the way people are actually going to react.

The whole aspect of what you might call acuity, people's ability to distinguish images and colour and so on, is different in moonlight from the theoretical answer. I'm not an expert in these things but there's no doubt – and I've experienced it myself – that one has a remarkably clear view of things in moonlight which, in theory, one should not be able to see. This is in moonlight of about 0.2 of a lux, whereas 20–25 lux is the normal requirement for reading. In four lux or five lux, which is twenty times moonlight, the colour and images that can be seen are remarkable.

It might also seem that there is not a lot of light. The size at which we expect to see things today, the concept of what the French call the *spectacle*, the event, means that we naturally, in spite of ourselves, overdo things, along the sort of argument of the philosophy of: 'well if we make it too big, we only need use half of it'. But of course that's not actually the nature of symmetry and balance. Invariably you want to use it all, and the only way you can use half

An Engineer Imagines

1992: theatre experiments on stage.

1992: computer drawing of new reflector prototypes intended to increase the amount and intensity of moonlight reaching the stage.

Summer 1992: Antares in the process of being hung, Pollux, Altair, Vega, Deneb and Polaris wait their turn. Timber structures are only half clad with flat mirror panels at this stage so that they don't get too heavy before being manhandled up the steps and hung in position on the scaffolding.

of it is to use a little bit in the middle. That is not very satisfactory either. We are fighting all our own bad habits all the time. We cannot seem to stop ourselves.

The other interesting thing is the quality of the light itself. Moonlight is almost perfect in its spectrum, the light of the moon is a very pure light except it has very little infra-red or ultra-violet, so it does not burn and it does not bleach. When we are looking at something by moonlight, we are actually looking at the true colours of the object we are lighting. Obviously concepts like true colours are strange concepts because everything has to be lit by something and you see only what is reflected. It is still an interesting psychological fact to know that, in abstract terms, one particular colour is true. If you could somehow find a way of transferring, say, a painting on to a negative without a bias when lit by moonlight, then you could get an amazing reproduction of the real colour because you would be eliminating most of the normal colour distortions. It's like innovation. It is necessary to get started and see what kind of distortion-free photographs one can achieve.

However, there is a real danger. Because these are very interesting problems, people are going to understand the potential of the reflector system. It will be important that the Centre does not become a research centre exploring the potential of the moonlight and the ways in which people react to it. That would be a negation of the whole thing. If the scientists go out there and measure the Full-Moon Theatre, it would become another instrument, another product of our industrial power. And in places where you can still think and talk, you just have to let things happen.

The director, Humbert Camerlo, calls it a *Théâtre d'Esquisse*, an experimental theatre, a theatre to 'try out new ideas' and experiment with interaction between performers and various art and science disciplines. It is by definition not a place for the general public but for performing art professionals. Its size is defined by the laboratory purpose, the minimum area of stage needed in relation to the nature of moonlight itself. For a large public theatre you would need absolutely massive reflectors way out of scale for this site. So this theatre can really only seat 200–250 people. What you can't do is say, 'turn up the moon a bit please.' You can do most other things. And so the other very important nature of the experience is that it is by necessity exclusive. It is excluded from our ability to build bigger and better, as we can with everything else today. You have an

Overall view of the amphitheatre with 1992 prototype reflectors in position.

The Full-Moon Theatre

aeroplane that flies 400 people across the Atlantic, and somebody says, 'Well, why not fly 700 people, it would be much more efficient.' And that's the internal contradiction – the more interesting it becomes the more people will want to participate. However, the moon only shines brightly enough for three or four nights a month around the time of the full moon and, as it is going to be cloudy through the winter period three or four times a year when the moon shines, the number of performances cannot be increased.

It is a device or an instrument which must be tested and be used by people themselves. Because of its very nature, people will have to find a place to make other moon theatres where they can become enthusiastic themselves. But this will always have to be at the same level, and always in the same way, because there is no other way you can do it. For all those kind of reasons I find it a passionate project. It is a project which you can immerse yourself in and allow it to generate its own momentum.

The idea is not mine, it is Humbert Camerlo's. The other thing one has to be very careful about with something like this is not to usurp the director's role. Our work does not for the moment overlap in the sense of our approach and expertise, so we are good partners. But Humbert is the energy of the place. Without him it would not exist.

There remain many technical problems to be resolved. These problems – the wind, the tracking systems for the reflectors, the light for musicians, to name but a few – are problems which can only be solved slowly with the means available at the Centre. Today we live in a world of instant solutions. Here that is not possible. The problem has got to be understood not just by Humbert and me but also, and principally, by the people permanently there.

Another special factor is that the project is not a problem we can solve quickly. It's not part of the modern misconception that we all have so little time to live, that if something is not done by tomorrow, then it is not worth doing. The Full-Moon Theatre is a project of many lifetimes.

Part of the difficulty at the moment is that the idea is so seductive. You need money to do anything and people only put money into something where they can see a result, so you have to produce some kind of result. Very few people are prepared for the result to be thirty years down the line. This is a basic dilemma. You have to go about it slowly, otherwise you are inevitably going to get it

1992: prototype reflectors all use flat mirror glass panels which can be adjusted both individually and as total frames which travel the independant curves of their top and bottom tracks. Dry stone wall tiers are modified to accomodate each new year's position requirements.

1992: audience below reflectors. Notice curving steel tracks along which the reflectors are moved manually at one meter per hour. The locus defined by top and bottom tracks ensures the correct tilt and focus.

Proposal for 1993 Kepler prototype using 1.3mm–1.6mm thick toughened mirror glass. This decrease in glass weight would better facilitate the movement of the reflector structure necessary to track the moon and stage action.

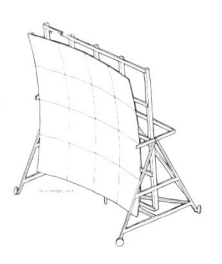

wrong. Each experience teaches probably one thing. So the development will have to be much more like the traditional growth of an indigenous culture and involve the modern ingredients which are part of our national capacity to think and to use all the thinking other people have done before.

The Full-Moon Theatre is a wonderful laboratory. All we can hope is that other people will be stimulated by it and not copy us but find their place and its problems, their myths, their breath. And eventually one could imagine maybe a dozen locations around the world where there would be moon theatres, all completely different but all generated by some reservoir of enthusiasm linked at a spiritual level. Gourgoubès will eventually become my spiritual home, that part of me which is not in Ireland. The Full-Moon Theatre is a project which is just beginning.

January 1993, midnight under a full moon. Writing with torch light.

Architecture in Movement

Watching sport has always played an important part in my life. I am fascinated by the parallels between the drama of sport and other events in real life. Flat racing has a strong affinity to architecture. As you watch a horse race unfold, it is like watching a city being built but it is taking place in real time. In the latter stages of many races, there is an instant in time when the whole future and outcome of the race is clear from the disposition of the horses. Each horse is as individual and separate as buildings would be in a square. The time that the race takes to run has its parallel in the time it takes to build the square. Time, after all, is no more than a mark or length on the unbounded totality of time which has no end. The ten seconds at the end of a great horse race and the 600 years to build Sienna have ultimately the same significance.

As I think of a race like the English 2,000 Guineas of 1984, of the moment when one elegant acceleration by Pat Eddery on El Gran Senor destroyed the field, I have the sense that I can touch and feel time. Thought of in this way, time becomes physical and part of the perceptible environment, like buildings and solid objects.

Horse-racing, at the highest level, is an unsurpassed study in motion of the shape of things to come. The finest horses can accelerate without any apparent change of effort and you know as you watch what the outcome will be. Frozen in the mind, moment after moment after moment.

Not all horses have the effortless elegance of El Gran Senor. (Another nice thing about horse-racing is the language and names of the horses. El Gran Senor, Chief Singer, Northern Dancer, Dancing Brave ... why can't we have names like that?)

As you watch a horse race, each successive instant tells a different story fully capturing the unfolding drama. I have never understood the scorn of the intelligentsia for activities like horse-racing and football. Football too has a wonderful sense of pattern when a moment of genius can change and wreck a carefully worked-out plan. Gaelic football, with its energy and its air of suppressed innocence, does not have the supreme artistry of Italian soccer, which in certain ways seeks to emulate the planned precision of American football, not something I enjoy.

Watching a horse race is like anything worth doing: it is something that has to be learned and studied, a frivolous activity maybe, but very rewarding. The step of the horses, the state of the ground, the going. Some years are a disaster: the way the horses gallop and

An Engineer Imagines

Aerial views of City of
London: 1939, 1947 and 1982.

move can change completely as the race unfolds second by second. The year Teenoso won the Derby, there was too much rain.

Everything is there in the picture. A truly three-dimensional and sculptural world in motion. Everyone should be required to learn about racing. It teaches you how to see when even the smallest event changes everything. A horse steps on an oversized clump of grass, swerves and loses momentum or accidentally confuses another horse's stride. Fragile, sensitive and elegant, the horses are supremely tuned machines which must be piloted with great sensitivity and awareness of their frailty. The jockeys transmits all through sensitive hands capable of influencing balance and stride by the way they sit, ready at any moment for the unexpected.

Gambling has for me nothing to do with the fascination I feel with the sport. It is the combination of two supreme and skilful athletes in harmony, pitted against others in the ever-changing unfolding world of the race. A new architecture instant by instant.

I have learned about horse-racing in tandem with Nemone, my Beaubourg daughter. Young eyes see so much more than older eyes. As I write a potentially great horse is waiting to test himself in the arena. It only happens once. Arazi won the Breeders Cup Mile for two-year-olds in incomparable style. One is lucky to be alive when a great racehorse comes along. They give us so much pleasure.

Architecture in Movement

Queen's Stand, Epsom.

Peter Rice at the Epsom Derby, 1992.

A sequence of seconds from the 1981 Extel Handicap shows the rhythmic style of Steve Cauthen on Number 13, 'Indian Trail'. Steve brings his mount with a smooth challenge on the outside to head 'Golden Flak' (Bruce Raymond) on the rails, and the hard-ridden 'Magikin' (Willie Carson). 'Bettyknowes' (Pat Eddery) is back in fourth place.

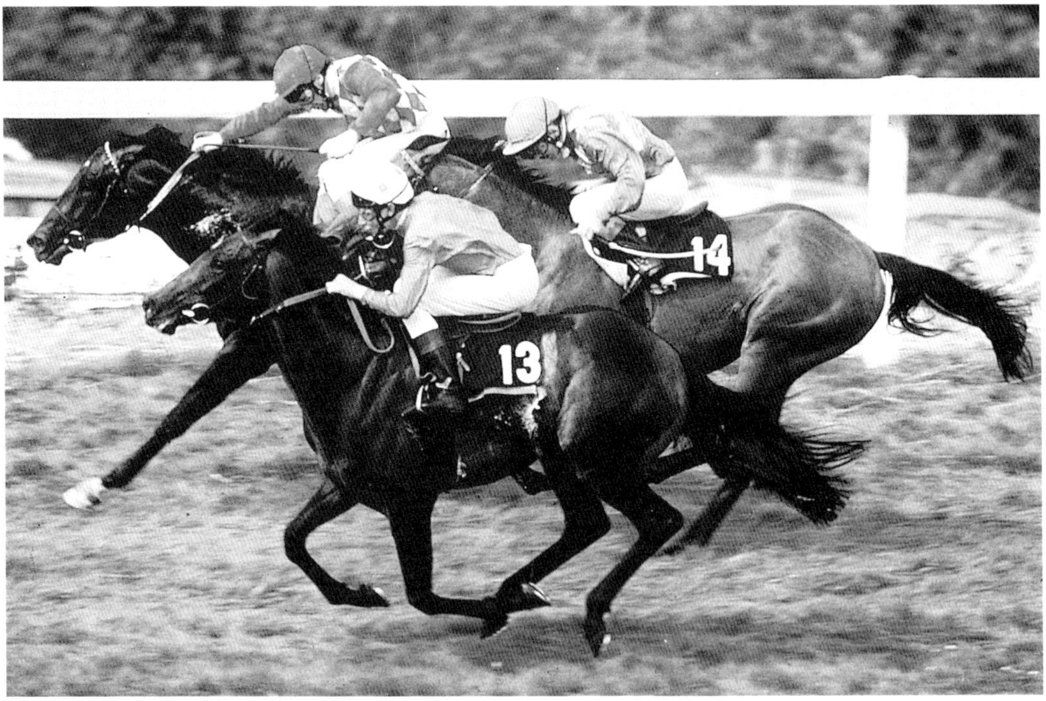

Epilogue

Stopping By Woods On A Snowy Evening

Whose woods these are I think I know.
His house is in the village, though;
He will not see me stopping here
To watch his woods fill up with snow.

My little horse must think it queer
To stop without a farmhouse near
Between the woods and frozen lake
The darkest evening of the year.

He gives his harness bells a shake
To ask if there is some mistake.
The only other sound's the sweep
Of easy wind and downy flake.

The woods are lovely, dark, and deep,
But I have promises to keep,
And miles to go before I sleep,
And miles to go before I sleep.

Robert Frost

Now that would be a challenge, to design a car as flexible as the horse.

Buildings and Projects

A Chronology

Except where otherwise noted, consulting engineer is Ove Arup & Partners.

1957

Sydney Opera House, Sydney, Australia
ARCHITECTS: Jørn Utzon (stages I and II), Hall Todd & Littlemore (stage III)

1967

Crucible Theatre, Sheffield
ARCHITECT:
Renton Howard Wood Associates

1969

Amberley Road Children's Home, London
ARCHITECT:
Renton Howard Wood Associates

Henrion Associates Consultancy Advice

1970

National Sports Centre, Crystal Palace, London
Proposals for a new stadium roof structure
ARCHITECT: Greater London Council Architects' Department

Circus 70, project, Victoria Embankment, London
Semi-permanent circular enclosed arena
ARCHITECT: Casson Conder & Partners

Arts Centre, Warwick University, Coventry
ARCHITECT: Renton Howard Wood Levin Partnership

Perspex spiral staircase, jeweller's shop, Jermyn Street, London
ARCHITECT:
Godfrey H & George P Grima

Super Grimentz Ski Village, Valais, Switzerland
New ski village
ARCHITECT:
Godfrey H & George P Grima

1971

Conference Centre, Mecca, Saudi Arabia
ARCHITECT: Rolf Gutbrod Architects and Professor Frei Otto

Special structures advice to Frei Otto and others on pneumatic and cable structures including 'The City in the Arctic'
ARCHITECT: Frei Otto

Centre Pompidou, Paris, France
ARCHITECT: Piano & Rogers

1972

World Trade Centre, London
Conversion of St Katharine's Dock House
ARCHITECT:
Renton Howard Wood Associates

1976

Jumbo jet hangar, Johannesburg, South Africa, project

1977

Shelter Span
Prefabricated building system
Peter Rice, consulting engineer

Fruili Housing, Italy
Free housing scheme
ARCHITECT:
Renzo Piano Building Workshop

Pilkington study
Prototype development of roofing units using glass-fibre reinforced cement
ARCHITECT: Richard Rogers & Partners

1978

Hammersmith Interchange, London
ARCHITECT: Foster Associates

Lloyd's of London Redevelopment, City of London
ARCHITECT: Richard Rogers & Partners

Industrialized construction system for Vibrocemento, Perugia, Italy
Piano & Rice

Il Rigo Quarter, Perugia, Italy
Housing prototype
Piano & Rice

Fiat VSS experimental vehicle, Turin, Italy
Piano & Rice

Fleetguard, Quimper, France
ARCHITECT: Richard Rogers & Partners

Patscentre, Princeton, New Jersey, USA
ARCHITECT: Richard Rogers & Partners

1979

Victoria Circus Shopping Centre, Southend on Sea, Essex
ARCHITECT: Alan Stanton

Educational television programme, RIA Television: The Open Site
Piano & Rice

An experiment in urban reconstruction for UNESCO, Otranto, Italy
Piano & Rice

1980

Design for Burano Island, Venice, Italy
Piano & Rice

Fabric roof canopy, Schlumberger Headquarters, Montrouge, France
ARCHITECT: Renzo Piano Atelier de Paris
CONSULTING ENGINEER: RFR in association with OAP

An Engineer Imagines

1981

Glass façades and central reception area roof, Cité des Sciences et de l'Industrie, La Villette, Paris, France
ARCHITECT: Adrien Fainsilber
CONSULTING ENGINEER: RFR

IBM Pavilion
ARCHITECT:
Renzo Piano Building Workshop

Stansted Airport Terminal Building, Stansted, Essex
ARCHITECT: Foster Associates

Menil Collection Museum, Houston, Texas, USA
ARCHITECT: Piano and Fitzgerald

1982

Alexandra Pavilion, London
ARCHITECT: Terry Farrell
CONSULTING ENGINEER:
Ove Arup & Partners
Peter Rice, consulting engineer for Shelter Span system

1983

Alton Towers, Alton, Staffordshire
Jet Star 2 Building
ARCHITECT: Griffin Jones Associates

Clifton Nurseries roof, Covent Garden, London
ARCHITECT: Terry Farrell

Pavilions, Tate Gallery, London
ARCHITECT: Alan Stanton

1984

122 St John Street, London
ARCHITECT: Eva Jiricna

Environment and motorway, Berlin, West Germany
Feasibility study for motorway acoustic protection system and solar heating for adjacent properties
ARCHITECT: Pascal Schöning

Ballsports stadium, Berlin, Germany
ARCHITECT: Christoph Langhof Architekten

1985

Emplacement, North Queensferry, Lothian, Scotland
Transformation of gun siting to residential workshop, project
ARCHITECT: Ian Ritchie Architects

Louvre, Paris, France
Design of steel structure to carry a glass roof over courtyards
ARCHITECT: I M Pei with Michel Macary

Lord's Mound Stand, London
ARCHITECT: Michael Hopkins & Partners

Aztec West Reception Building, Bristol
ARCHITECT: Michael Hopkins Architects

Roy Square, Narrow Street, London
ARCHITECT: Ian Ritchie Architects

Atrium roof, Conflans, Saint Honorine, France
ARCHITECT: Valode et Pistre

Fresco restoration for Menil Collection, Houston, Texas, USA
Art restorer: Laurence Morocco

1986

Fabric canopy, St Louis/Basle, France
ARCHITECT:
Aéroports de Paris/Paul Andreu
CONSULTING ENGINEERS: RFR in association with Ove Arup & Partners

Nuage Léger, Tête Défense, La Défense, Paris, France
ARCHITECT: J O Spreckelsen and Aéroports de Paris/Paul Andreu
CONSULTING ENGINEERS: RFR in association with Ove Arup & Partners

Nuage Parvis, La Défense, Paris, France
ARCHITECT: J O Spreckelsen and Aéroports de Paris/Paul Andreu
CONSULTING ENGINEERS: RFR

Passerelle Lintas, Paris, France
ARCHITECT: Marc Held
CONSULTING ENGINEERS: RFR

Canopies, Parc de La Villette, Paris, France
ARCHITECT: Bernard Tschumi
CONSULTING ENGINEERS: RFR

Central House, Whitechapel High Street, London
ARCHITECT: Ian Ritchie Architects

Football Stadium, Bari, Italy
ARCHITECT: Renzo Piano Building Workshop

IBM 'Ladybird' Travelling Exhibition, Italy
ARCHITECT:
Renzo Piano Building Workshop

Apartment for John Young, London
ARCHITECT: John Young

Opéra, Bastille, Paris, France
Studies for acoustic ceiling
ARCHITECT: Carlos Ott
CONSULTING ENGINEERS: RFR

Centre Industriel, Epone, France
Steel warehouse hypermarket structure
ARCHITECT: Richard Rogers & Partners

Centre Industriel, St Herblain, Nantes, Loire-Atlantique, France
Steel warehouse hypermarket structure
ARCHITECT: Richard Rogers & Partners

Buildings and Projects
A Chronology

Floating restaurant, Jubilee Gardens, London
ARCHITECT: Richard Rogers & Partners

European Synchotron Radiation Facility, Grenoble, France
ARCHITECT: Renzo Piano Atelier de Paris

1987

Parc Citroën Cévennes, Greenhouses, Paris, France
ARCHITECT: Patrick Berger
CONSULTING ENGINEER: RFR

Passerelles Front de Seine, Paris, France
DESIGN AND CONSULTING ENGINEER: RFR

Fabric roof covering, Château de Falaise, Normandy, France
ARCHITECT: Decaris
CONSULTING ENGINEER: RFR

Glazed façade, Musée des Beaux Arts de Clermont-Ferrand, France
ARCHITECT: A Fainsilber and Gaillard
CONSULTING ENGINEER: RFR

Multi-purpose hall, Nancy, France
Design for 70-metre-span cable-braced roof
ARCHITECT: Foster Associates

58-metre motor yacht
Computer-aided design work for the stability of the yacht
NAVAL ARCHITECT: Martin Francis

Sports Hall, Ravenna, Italy
ARCHITECT: Renzo Piano Building Workshop

Competition for aircraft hangars, Abu Dhabi, United Arab Emirates
ARCHITECTS: Aéroports de Paris/Paul Andreu

Azabu and Tomigaya Structure, Tokyo, Japan
ARCHITECT: Zaha Hadid

Office/apartment block, Lecce, Italy
ARCHITECT: Renzo Piano Building Workshop

Centre Industriel, Massy, Essonne, France
ARCHITECT: Richard Rogers & Partners

UNESCO Laboratory/Building Workshop, Genoa, Italy
ARCHITECT: Renzo Piano Building Workshop

Atrium, offices for Bull, Avenue Gambetta, Paris, France
ARCHITECT: Valode et Pistre
CONSULTING ENGINEERS: RFR in association with Ove Arup & Partners

1988

La Grande Nef, Tête Défense, Paris, France
ARCHITECT: Jean-Pierre Buffi
CONSULTING ENGINEER: RFR

TGV/RER Charles de Gaulle, Roissy, France
ARCHITECT: Aéroports de Paris/Paul Andreu
CONSULTING ENGINEER: RFR

Les Tours de la Liberté, Paris, France
ARCHITECT: Hennin et Normier
CONSULTING ENGINEERS: RFR in association with Ove Arup & Partners

Façade of the BPOA, Rennes, France
ARCHITECT: O Decq and B Cornette
CONSULTING ENGINEER: RFR

Franconville, France
ARCHITECT: Cuno Brullmann, Arnaud Fougeras Lavergnolle Architects

Chur, Switzerland
Glazed roof canopy, bus/rail station
ARCHITECT: Robert Obrist and Richard Brosi
CONSULTING ENGINEERS: Ove Arup & Partners in association with RFR

Piazza, Stag Place Site B, London
ARCHITECT: Richard Rogers Partnership

Sistiana, Italy
Studies for conversion of disused quarry and sea frontages into resort
ARCHITECT: Renzo Piano Building Workshop

Crown Princess liner, Italy
Design of the superstructure for the refurbishment of a liner
ARCHITECT: Renzo Piano Building Workshop

Museum of Contemporary Art, Bordeaux, France
ARCHITECT: Valode et Pistre

Centre Industriel, Pontoise, Val d'Oise, France
ARCHITECT: Richard Rogers Partnership

Pearl of Dubai, Dubai, United Arab Emirates
20-metre sphere and museum project
ARCHITECT: Ian Ritchie Architects

Harbour Museum, Newport Beach, California, USA
ARCHITECT: Renzo Piano Building Workshop

RAF Northolt, London
New air terminal
ARCHITECT: Property Services Agency

Kansai International Airport Terminal, Japan
ARCHITECT: Renzo Piano Building Workshop

Europe House, World Trade Centre, London
ARCHITECT: Richard Rogers Partnership

Marseilles Airport Terminal Building, France
ARCHITECT: Richard Rogers Partnership

Grand Louvre, Paris, France
Observation pavilions
ARCHITECT: Michael Dowd

National Portrait Gallery, London
ARCHITECT: Stanton Williams

Bercy 2, France
Roof system for shopping centre
ARCHITECT: Renzo Piano Atelier de Paris

Studies for Discoveryland Mountain,
Eurodisney, France
ARCHITECT: Disney Imagineering
CONSULTING ENGINEERS: Ove Arup &
Partners in association with RFR

Competition for Patinoire d'Albertville,
France
Covered skating rink for Winter
Olympics
ARCHITECT: Adrien Fainsilber
CONSULTING ENGINEERS: Ove Arup &
Partners in association with RFR

Full-Moon Theatre, Provence, France
DESIGNER AND THEATRE DIRECTOR:
Humbert Camerlo
CONSULTING ENGINEERS:
Ove Arup & Partners and RFR

Hull design, Staquote British Defender
yacht, Whitbread Round-the-World
Race
NAVAL ARCHITECT: Martin Francis

Porte Monumentale, Paris, France
ARCHITECT: A Zublena
CONSULTING ENGINEER: RFR

1989

Glass canopy, Verdun, France
ARCHITECT: Pierre Colboc
CONSULTING ENGINEER: RFR

Sloping glazed façade and roof, Shell,
Rueil Malmaison, France
ARCHITECT: Valode et Pistre
CONSULTING ENGINEERS: Ove Arup and
Partners in association with RFR

Passerelle Est/Ouest, Galeries
Nord/Sud and bridge, Parc de la
Villette, Paris, France
ARCHITECT: Bernard Tschumi
CONSULTING ENGINEER: RFR

L'Oréal factory, Aulnay, France
ARCHITECT: Valode et Pistre

Pavilion of the Future, Expo '92, Seville,
Spain
ARCHITECT: Martorell Bohigas Mackay

Industrial buildings, Tomaya, Japan
ARCHITECT: Sugimur

Queen's Stand, Epsom Racecourse,
Surrey
ARCHITECT: Richard Horden Associates

Toronto Opera House, Canada
ARCHITECT: Moshe Safdie & Associates

Il Grande Bigo, Colombo 500, Expo '92,
Genoa, Italy
ARCHITECT:
Renzo Piano Building Workshop

Mobile sculpture, Colombo 500, Expo
'92, Genoa, Italy
ARTIST: Susumo Shingu
ARCHITECT:
Renzo Piano Building Workshop

Pharmacy, Boves, France
ARCHITECT: Ian Ritchie Architects

Helios VII
Feasibility study for summer house
based on solar sculpture

Spiders' webs
Research project
ZOOLOGIST: Dr Fritz Vollrath

Office building, Stockley Park, London
ARCHITECT: Ian Ritchie Architects

Kurfürstendamm, Berlin, Germany
Façade study
ARCHITECT: Zaha Hadid, Stefan Schroth
CONSULTING ENGINEERS:
Ove Arup & Partners and RFR

Mitsubishi Tokyo Forum competition,
Japan
ARCHITECT: Richard Rogers Partnership

Ecology Gallery, Natural History
Museum, London
ARCHITECT: Ian Ritchie Architects

La Villette serre study, 'Green' Spiral,
Musée des Sciences et de l'Industrie, La
Villette, Paris, France
ARCHITECT: Kathryn Gustafson

Utsurohi, La Défense, Paris, France
Sculpture
ARTIST: Iyo Miyawaki

1990

Station Square competition,
Oberhausen, Germany
DESIGN ENGINEERING: RFR

Canopy and façades for CNIT, Paris,
France
CONSULTING ENGINEER: RFR

Campanile, Place d'Italie, Paris, France
Tower structure with sculpture
ARCHITECT: Kenzo Tange
SCULPTOR: Thierry Vidé

Atrium roof, Grand Ecran, Place
d'Italie, Paris, France
ARCHITECT: Kenzo Tange
CONSULTING ENGINEER: RFR

IMAX cinema and leisure building,
Liverpool
ARCHITECT: Richard Rogers Partnership

Groningen Museum, Groningen,
Holland
Preliminary study of structure of roof
ARCHITECT: Alessandro Mendini
ARTIST: Frank Stella

Padre Pio Pilgrimage Church, St
Giovanni Rotondo, Puglia, Italy
ARCHITECT:
Renzo Piano Building Workshop

Buildings and Projects
A Chronology

Centre Culturel de la Pierre Plantée
Bibliothèque, Vitrolles, France
ARCHITECT: Ian Ritchie Architects

Development of glazing system with
Asahi Glass, Japan
DESIGN: RFR and Ove Arup & Partners

1991

Aéroport Charles de Gaulle, Terminal 3,
Roissy, France
ARCHITECT:
Aéroports de Paris/Paul Andreu
CONSULTING ENGINEER: RFR

Atrium, Centre D'Art Contemporain,
Luxembourg
ARCHITECT: I M Pei
CONSULTING ENGINEER: RFR

Lamp post, Esch sur Alzette,
Luxembourg
ARCHITECT AND TOWN PLANNER:
Professor Sieverts
DESIGN AND ENGINEERING: RFR

Pyramide Inversée, Grand Louvre,
Paris, France
Inverted glass pyramid sculpture
suspended over underground public
circulation area at museum
ARCHITECT: I M Pei
CONSULTING ENGINEER: RFR

Japan Bridge, Paris, France
ARCHITECT: Kisho Kurokawa
CONSULTING ENGINEERS: RFR in association
with Ove Arup & Partners

Swimming pool, Levallois Perret,
France
ARCHITECT: Cuno Brullmann
CONSULTING ENGINEER: RFR

Rampe R4/T4, Parc de la Villette, Paris,
France
ARCHITECT: Bernard Tschumi
CONSULTING ENGINEER: RFR

Passerelle, Mantes-la-Jolie, France
ARCHITECT: Michel Macary
CONSULTING ENGINEER: RFR

Ligne Météor, Paris, France
Stations on new Metro line from Gare
St Lazare to Place d'Italie
ARCHITECT: Bernard Kohn
CONSULTING ENGINEER: RFR

Aérogare de Luxembourg
ARCHITECT: Bohdan Paczowski
CONSULTING ENGINEER: RFR

Hôtel Module d'Echanges, Roissy,
France
ARCHITECT:
Aéroports de Paris/Paul Andreu
CONSULTING ENGINEER: RFR

Cathédrale Nôtre Dame de la Treille,
Lille, France
ARCHITECT: Pierre-Louis Carlier, with
ARTIST L Kinjo
CONSULTING ENGINEER: RFR

River bus station project, Lyon, France
ARCHITECT:
C Charignon, D Cornilliat and P Levy
CONSULTING ENGINEER: RFR

Glass façades, Renault Technocentre,
Guyancourt, France
ARCHITECT: Valode et Pistre
ENGINEERING: Ove Arup & Partners in
association with RFR

TGV Station, Lille, France
Roof design
ARCHITECT: SNCF Jean-Marie Duthilleul
CONSULTING ENGINEERS: Ove Arup &
Partners in association with RFR

Museum of the Moving Image
demountable tent, London
ARCHITECT: Future Systems

Brau & Brunnen Tower, Berlin,
Germany
ARCHITECT: Richard Rogers & Partners

Atrium glazing, 50 Avenue Montaigne,
Paris, France
ARCHITECT: O Vidal
CONSULTING ENGINEER: RFR

Grand Louvre, Louvre Museum, Paris,
France
Natural lighting system for new
museum
ARCHITECT: Pei Cobb Freed & Partners

Ajaccio Airport, Corsica
ARCHITECT:
Aéroports de Paris/Paul Andreu
CONSULTING ENGINEER: RFR

Glazing for Digital Headquarters,
Geneva, Switzerland
Technical assistance to SIV, glass
manufacturers, Italy
CONSULTING ENGINEER: RFR

Fabric sculpture, Sainsbury's,
Plymouth, Devon
ARCHITECT: Dixon Jones

1992

Façade for the extension of the Palais de
Congrès, Paris, France
ARCHITECT: Olivier Clement Cacoub
CONSULTING ENGINEER: RFR

Passerelle, Levallois-Perret, France
ARCHITECT: Henri Caubel
CONSULTING ENGINEER: RFR

Atrium, Museum of Modern Art,
Strasbourg, France
ARCHITECT: Adrien Fainsilber
CONSULTING ENGINEER: RFR

Cultural Centre, Albert, France
ARCHITECT: Ian Ritchie Architects

Western Morning News, Plymouth
ARCHITECT: Nicholas Grimshaw and
Partners

Bibliography

WORKS BY PETER RICE

P Rice (1971), 'Notes on the Design of Cable Roofs', *Arup Journal*, vol. 6, no. 4, pp. 6–10

E F Happold and P Rice (1973), 'Introduction', *Arup Journal*, Centre Beaubourg special issue, vol. 8, no. 2, June, pp. 2–3

P Rice and L Grut (1975), 'Main Structural Framework of the Beaubourg Centre, Paris, *Acier Stahl Steel*, vol. 40, no. 9, September, pp. 297–309

P Rice and R Peirce (1976), 'The Barrettes of Centre Georges Pompidou', South African Institution of Civil Engineers, Fifth Quinquennial Regional Convention, Stellenbosch University, 28–30 September (Innovations in Civil Engineering conference paper)

P Rice and A Day (1977), 'Lagoon Barriers at Venice', *Arup Journal*, vol. 12, no. 2, pp. 21–22

P Rice (1977), 'Materials in Use: Fire Protection and Maintenance at the Centre Pompidou, *RIBA Journal*, vol. 84, no. 11, November, pp. 476–477

P B Ahm, F G Clarke, E L Grut and P Rice (1979, 1980), 'Design and Construction of the Centre National d'Art et de Culture Georges Pompidou', *ICE Proceedings*, part 1, no. 66, pp 557–593 (November 1979), part 1, no. 68, pp. 499–505 (August 1980)

P Rice (1980), 'The Structural and Geometric Characteristics of Lightweight Structures', Ove Arup Partnership: Arup Partnerships' Seminar: Lightweight Structures (March) 8pp

P Rice (1980), 'Lightweight Structures: Introduction', *Arup Journal*, vol. 15, no. 3, October, pp. 2–5

P Rice (1981), 'Long Spans and Soft Skins', *Consulting Engineer*, vol. 45, no. 7, pp. 10–12

P Rice, T Barker, A Guthrie and N Noble (1983), 'The Menil Collection, Houston Texas', *Arup Journal*, vol. 18, no. 1, pp. 2–7

J Young and J Thornton (1984), 'Design for Better Assembly: (5) Case Study: Rogers' and Arup's', *The Architects' Journal*, vol. 180, no. 36, 5 September, pp. 87–94

A A Tassel and P Rice (1985), 'Master of the Hi-tech Style. Britain's Peter Rice', *SA Construction World*, April, pp. 54–61

P Rice (1986), 'Rogers Revolution', *Building Design*, no. 807, 10 October, pp. 32–33

P Rice and J A Thornton (1986), 'Lloyd's Redevelopment', *Structural Engineer*, vol. 64A, no. 10, October, pp. 265–281

A Day, T Haslett, T Carfrae and P Rice (1986), 'Buckling and non-linear behaviour of space frames', International Conference on Lightweight Structures in Architecture, Sydney, no. 1, August (LSA 86), pp. 775–782

P Rice (1987), 'Menil Collection Museum Roof: Evolving the Form', *Arup Journal*, vol. 22, no. 2, Summer, pp. 2–5

P Rice (1987), 'The Controlled Energy of Renzo Piano', *Renzo Piano: The Process of Architecture*, exhibition catalogue, 1987 (9H Gallery, London)

P Rice, J A Thornton, A Lenczner (1988), 'Cable Stayed Roofs for Shopping Centres at Nantes and Epone', *Structural Engineering Review*, no. 1, pp. 133–140

Interview with Peter Rice, *Le Moniteur*, no. 4 453, 31 March 1989, pp. 46–47

P Rice (1989), 'A Celebration of the Life and Work of Ove Arup', *RSA Journal*, June, pp. 425–437, reprinted in *Arup Journal*, vol. 25, no. 1, Spring 1990, pp. 43–47

P Rice (1989), 'Building as Craft, Building as Industry',*Guggenheim Museum Proceedings*, pp 87–89 (New York)

P Rice (1990) 'Constructive Intelligence', *Arch +*, no. 102, January, pp. 37–52

C Gabato and P Rice (1990), 'Bari', *Architecktur Aktuell*, vol. 24, no. 139, October, pp. 82–85

P Rice (1990), 'Unstable Structures', *Columbia Documents of Architecture and Theory*, vol. 1 (Rizzoli, New York), pp. 71–90

P Rice and H Dutton (1990), *Le Verre Structurel* (Editions du Moniteur, Paris)

P Rice, A Lenczner, T Carfrae, and A Sedgwick (1990), 'The San Nicola Stadium, Bari', *Arup Journal*, vol. 25, no. 3, Autumn, pp. 3–8

P Rice, A Lenczner and T Carfrae (1991), 'The San Nicola Stadium, Bari', *Steel Construction Today*, vol 5, no. 4, July, pp. 157–160

P Rice (1991), 'Practice and Europe', *The Structural Engineer*, vol. 69, no. 23, 3 December, pp. 400–401

P Rice (1991), 'Architecture Now, Arcam Pocket', chapter: 'Peter Rice –Great Britain'. *Architectura & Natura*, pp 127–128

P Rice (1991), 'Materials and Design Using Timber as a Model', International Timber Engineering Conference, London

P Rice (1991), 'Menil Collection Museum Roof: evolving the form', Offramp, vol. 1, no. 4, pp. 117–119

P Rice (1992), 'Dilemma of Technology' in A Tzonis and L Lefaivre, Architecture in Europe since 1968: Memory and Intervention (Thames & Hudson, London), pp 36–41

P Rice (1992), Speech at 1992 RIBA Gold Medal Presentation, RIBA Journal, September

ARTICLES ABOUT PETER RICE

(1984), 'Interview with Architect John Young and Engineers Peter Rice and John Thornton', The Architect's Journal, vol. 180, no. 361, 5 September, pp. 87–94

(1985), 'The Nature of Materials: Interview with Peter Rice', Architecture d'Aujourd'hui, no. 237, February, pp. 10–17

(1986), 'La Villette: Cité des Sciences et de l'Industrie', Techniques et Architecture, no. 364, February–March, pp. 60–129

(1987), 'Il Punto di Vista di Peter Rice' ('An Engineer's View'), L'Arca, no. 5, April, pp. 70–75

(1987), 'Soaring Aspirations: Peter Rice talks about his admiration for the Gothic masons', Royal Academy Magazine, no. 17, Winter, pp. 29-30

(1987), 'Stratégie de l'Araignée', Architecture d'Aujourd'hui, no. 252, September, pp. 78–79 (proposal for tensile structure at La Défense)

(1989), 'Pei As You Learn: Exhibition Pavilion, The Louvre, Paris – Architects: Michael Dowd with Peter Rice', Building Design, 17 February, pp. 28–29

(1991), Building Arts Forum, New York, Bridging the Gap: Rethinking the Relationship of Architect and Engineer (Van Nostrand Reinhold, New York)

(1992), Conversation with Peter Rice: Exposed Structures in Building Design (McGraw-Hill, New York)

D D Boles (1987), 'The Point of No Return', Progressive Architecture, July 1987, pp. 94–97 (progress report on Parc de la Villette)

P Buchanan (1983), 'Patscenter, Princeton', Architectural Review, vol. 174, no. 1,037: pp. 43-47 (appraisal of the scheme, with comments on the design concept by Peter Rice)

D Chipperfield (1988), 'Tecnologia i Arquitectura = Technology and Architecture', Quaderns d'Arquitectura i Urbanisme, no. 178, July–September, pp. 120–125 (interview with Norman Foster with excerpt from interview with Peter Rice)

P Cook (1987), 'Glass Link', Blueprint, no. 37, May, pp. 26–27 (courtyard pedestrian bridge, Quai de la Megisserie, Paris)

P Cook (1988), 'Passerella in Vetro a Parigi', Abacus, vol. 4, no. 14, June–July, pp. 5, 10, 66–71

'P D' (1984), 'Space Wagner: Project for a Re-locatable Opera House', Architectural Review, vol. 175, no. 1,046, 1984, pp. 54–56

P Davey (1987), 'Operation Overlord's', Architectural Review, vol. 82, no. 1,087, September, pp. 40–49 (Lord's Cricket Stand, London)

C Ellis (1987), 'Unused Abattoir Becomes a Bristling Science Museum', Architecture, September, pp. 85–87 (the City of Science and Industry at La Villette)

J Ferrier, Jacques and A Lavalou (1990), 'Figures Brittaniques: Peter Rice', Architecture d'Aujourd'hui, no. 267, February, pp. 114–121 (interview and three projects: Tête-Défense, Lintas Footbridge, Bari Stadium)

O Fillion (1989), 'Peter Rice, la Voie Etroite des Concepteurs', Moniteur, no. 4,453, 31 March, pp. 46–47

Bibliography

T Fisher (1989), 'A Non-Unified Field Theory', *Progressive Architecture*, November, pp. 65–73, 135 (Parc de la Villette, with details of canopy)

P Goulet (1985), 'La Nature des Materiaux: de l'Acier Moulé au Polycarbonate – Entrentien avec Peter Rice', *Architecture d'Aujourd'hui*, no. 237, February, pp. 10–17

S Grôâk (1984), 'Engineers as Pioneers', *The Architects' Journal*, vol. 180, no. 45, 7 November, pp. 48–49 (review of lectures given by Tony Hunt and Peter Rice at the RIBA)

L Knobel (1983), 'Striking Through the Mask', *Domus*, no. 637, March, pp. 10–23 (high-tech, including the work of Peter Rice)

L Knobel (1982), 'Tate Pavilions', *Architectural Review*, vol. 172, no. 1,027, September, pp. 70–73

P C Papademetriou (1987), 'The Responsive Box', *Progressive Architecture*, May, pp. 87–97 (the Menil Collection)

M Pawley (1989) 'The Secret Life of the Engineers', *Blueprint*, no. 55, March, pp. 34–36 (Frank Newby, Peter Rice, Tony Hunt)

A Pélissier (1986), 'Peter Rice à la Villette', *Techniques et Architecture*, no. 364, February–March, pp. 134–135

R Piano, R MacCormac, R Rogers (1992), 'Royal Gold Medal Address', *RIBA Journal*, vol. 99, September, pp. 26–29

V Picon-Lefebvre (1990), 'Travaux d'ingenieurs', *Moniteur architecture*, no. 15, October, pp. 40–49

J Pink (1992), 'Future Shock', *Architectural Review*, vol. 190, no. 1144, June, pp. 53–54 (Seville Expo)

A Plowman (1985), 'Pick of the "Cover" Storeys?', *Glass Age*, April (projects using glass, including Lloyd's)

L Relph-Knight (1984), 'Paris, Texas …', *Building Design*, 26 October, pp. 18–19

D Sudjic (1980), 'High Tech with a Human Face', *The Architects' Journal*, vol. 172, no. 28, 9 July, pp. 74–75 (Piano & Rice's mobile workshop for Otranto)

B Suner (1989), 'Ossature des Nuages', *Architecture d'Aujourd'hui*, no. 265, October 1989, pp. 196, 198–211 (the suspended tent structure inside the Cube of the Grande Arche de la Défense)

B Waters (1981), 'Setting the Sails', *Building*, vol. 241, no. 37, pp. 32–34 (the Alexandra Palace tent)

OBITUARIES

(1992), 'Peter Rice: tributes to a great structural engineer', *The Architects' Journal*, 4 November

(1992), 'Obituary', *RIBA Journal*, vol. 99, December, p. 64

(1992), 'Peter Rice', *The Times*, 7 November

M Dowd (1992), 'Peter Rice', *Design Week*, 6 November, p. 3

J Ferrier (1992), 'Peter Rice le grand ingénieur high tech est décédé', *Architecture d'aujourd'hui*, no. 284, December, pp. 26–28

J Glancey (1992), 'Peter Rice', *The Independent*, 29 November

D Gruber (1992), 'Reflections on a Consummate Artisan', *Progressive Architecture*, December, pp. 84–87

M Pawley (1992), 'Ricordo di Peter Rice', *Casabella*, no. 596, December, p. 21

I Ritchie (1992), 'Peter Rice 1935–1992', *Building Design*, 30 October

C H Stanton (1992), 'Peter Rice, a Retrospective', *New Civil Engineer*, 12 November, pp. 16–17

J Zunz (1992), 'The Master Builder', *Guardian*, 28 October

CATALOGUES, FILMS

P Rice (1986), *Exploring the Boundaries of Design*, library of tape slide lectures; Peter Rice talks about his early career and how materials have influenced his projects (Pidgeon Audio Visual 15/8610, London)

Exploring Materials: the Work of Peter Rice, Royal Gold Medallist 1992, catalogue to accompany exhibition, prepared by B-A Campbell and J McMinn with the Ove Arup Partnership

P Rice, *'La Cité en Luminaire'*, La Villette; video made by RFR, explains the design and construction of the Grandes Serres

P Rice, *The Structural Engineer*, video of lecture given on 17 January 1990 as part of the 'Engineers in Design' series, introduced by Bob Emmerson. Peter Rice discusses the role of the structural engineer, the nature of materials and of stability, and the mathematics of chaos

The Late Show, BBC TV, 1992; profile of Peter Rice on winning the RIBA Gold medal. Features major projects and tributes from Michael Hopkins, Sir Richard Rogers, Nicholas Grimshaw and Frank Stella

Illustration Credits

Illustrations are credited from left to right and from top to bottom of each page. OAP = Ove Arup & Partners

Cover Humbert Camerlo
Frontispiece Michel Denancé

COLOUR PLATES
Wild flowers J Hyett;
Sydney Opera House Harry Sowden;
Centre Pompidou Bernard Vincent;
Nuage Cube OAP;
La Villette RFR;
Pavilion of the Future, Seville Alistair Lenczner/OAP;
Gourgoubès Humbert Camerlo

Between pp. 56 and 57
Peter Mackinven/OAP
Between pp. 72 and 73 Zaha M Hadid
Between pp. 80 and 81 Jay Maisel
Between pp. 136 and 137
Frank Stella/Steven Sloman

FINAL COLOUR SECTION
Street lights Kira Leroy, Antoine Baz; RFR;
Esch Arbed; Robert Champagne;
Japan Bridge Kate Purver; Viry;
Lyon RFR;
Kansai Renzo Piano Building Workshop – Japan/OAP;
Lille SNCF; Michel Denancé;
Roissy Ana Maksimiuk;
Chur Hugh Dutton;
Pyramide Inversée Peter Rice sketchbooks; RFR;
CNIT Michel Denancé;
Padre Pio Pilgrimage Church Renzo Piano Building Workshop/OAP;
Fresco A C Cooper/The Menil Foundation, Houston

p. 24 Renzo Piano Building Workshop
p. 26 Archigram Archive; OAP; Shunji Ishida
p. 27 all pictures, Piano & Rogers
p. 28 Renzo Piano Building Workshop
p. 29 Masao Arai (Shinkenchiku-sha); Dr Bill Addis; Dr Bill Addis
p. 31 OAP

p. 32 Karl Mohringer; Matthias Kutterer; Annalie Riches; Hugh Dutton
p. 33 all pictures, OAP
p. 34 Gianni Berengo Gardin
p. 38 Bernard Vincent
p. 39 Gianni Berengo Gardin
p. 40 Gianni Berengo Gardin
p. 41 Bernard Vincent
p. 42 Bernard Vincent
p. 43 Bernard Vincent
p. 46 Shunji Ishida; Gianni Berengo Gardin
p. 58 Jørn Utzon
p. 60 Jørn Utzon; News Limited Sydney; Jørn Utzon
p. 61 Sir Jack Zunz/OAP
p. 62 Jørn Utzon; Jørn Utzon; Jørn Utzon; Max Dupain; OAP
p. 63 all pictures, OAP
p. 64 all pictures, Harry Sowden
p. 65 all pictures, Harry Sowden
p. 66 all pictures, Harry Sowden
p. 70 Michael Peto
p. 71 The National Motor Museum at Beaulieu
p. 72 OAP; OAP; Alain Goustard – Archipress
p. 74 Povl Ahm/OAP; Travers Morgan Ltd and The New Civil Engineer Magazine
p. 76 Shakespeare Centre Library, Joe Cocks Studio Collection
p. 77 B Mancia, F Bodmer, FBM Studio; John James Collection, Institute of Civil Engineers; John James Collection, Institute of Civil Engineers
p. 78 Robert Champagne; Gianni Berengo Gardin
p. 79 OAP; Renzo Piano Building Workshop
p. 81 SPADEM
p. 82 all pictures, SPADEM
p. 83 all pictures, SPADEM
p. 84 all pictures, SPADEM
p. 85 Michel Denancé; RFR; RFR; SPADEM
p. 86 Ben Smusz
p. 88 Renzo Piano Building Workshop; Peter Rice sketchbooks; Renzo Piano Building Workshop; Peter Rice sketchbooks
p. 89 all pictures, Renzo Piano Building Workshop

p. 90 David Crossley; Shunji Ishida; David Crossley
p. 91 Renzo Piano Building Workshop; Gianni Berengo Gardin; Gianni Berengo Gardin; Richard Bryant/ARCAID
p. 92 David Crossley; Renzo Piano Building Workshop; Gianni Berengo Gardin; OAP; OAP; OAP; Jane Reynolds; David Crossley
p. 93 Richard Bryant/ARCAID; OAP
p. 94 Richard Bryant/ARCAID
p. 95 Jay Maisel
p. 96 Professor Frei Otto; Professor Frei Otto; Professor Frei Otto; OAP
p. 97 Michel Denancé; OAP; Renzo Piano Building Workshop; Alistair Lenczner/OAP
p. 98 Peter Mackinven/OAP
p. 99 Future Systems; Future Systems; Geoffrey Beeckman
p. 100 J O Spreckelsen; Koit; OAP
p. 101 Peter Rice Sketchbooks; OAP
p. 102 RFR; RFR; RFR/ADP
p. 103 all pictures, RFR
p. 104 Alan Stanton; Alan Stanton; unknown; unknown
p. 105 Mount Holyoke College Art Museum, South Hadley, Massachusetts; RFR; RFR; RFR; RFR; RFR
p. 106 Gianni Berengo Gardin; Shunji Ishida; Emanuela Minetti; Barbara-Ann Campbell; Ana Maksimiuk
p. 107 Gianni Berengo Gardin
p. 108 Professor Kurt Ackermann; Renzo Piano Building Workshop; Shunji Ishida; Gianni Berengo Gardin; Robert Kinch/OAP; Gianni Berengo Gardin; Gianni Berengo Gardin; Gianni Berengo Gardin
p. 109 Renzo Piano Building Workshop; Professor Kurt Ackermann; Robert Kinch/OAP; Gianni Berengo Gardin; Gianni Berengo Gardin; Professor Kurt Ackermann; Gianni Berengo Gardin; Gianni Berengo Gardin; Gianni Berengo Gardin
p. 110 Stéphane Couturier – Archipress
p. 111 Alain Goustard – Archipress; RFR; RFR; RFR; RFR
p. 112 RFR; RFR; RFR; OAP; Barbara-Ann Campbell
p. 113 OAP; Michel Denancé - Archipress

p. 114 Harry Sowden/OAP; OAP
p. 115 all pictures, OAP
p. 116 Renzo Piano Building Workshop; Annalie Riches/RFR; OAP
p. 117 Richard Rogers Partnership/OAP; OAP; Harry Sowden/OAP; Richard Rogers Partnership/OAP; Harry Sowden/OAP
p. 118 David Glover/OAP
p. 119 Alistair Lenczner/OAP
p. 120 Peter Rice sketchbooks
p. 121 Bruce Danziger/OAP; Bruce Danziger/OAP; OAP
p. 122 all pictures, OAP
p. 123 OAP; Professor Jacques Heyman; OAP; OAP
p. 124 all pictures, OAP
p. 125 all pictures, OAP
p. 128 all pictures, Harry Sowden/OAP
p. 130 all pictures, Otto Baitz
p. 131 OAP
p. 134 Souhei Imamura/Architectural Association Slide Library
p. 136 Shunji Ishida; Fiat Auto
p. 137 OAP; Fiat Auto
p. 138 Renzo Piano Building Workshop; Renzo Piano Building Workshop; Renzo Piano Building Workshop; Fiat Auto; Shunji Ishida; Renzo Piano Building Workshop
p. 140 Renzo Piano Building Workshop; Shunji Ishida; Renzo Piano Building Workshop; Renzo Piano Building Workshop
p. 143 Fiat Auto
p. 144 Stéphane Couturier – Archipress
p. 146 RFR; Alain Goustard – Archipress
p. 147 Luc Boegley – Archipress; Matthias Kutterer; Matthias Kutterer
p. 148 Humbert Camerlo
p. 150 Peter Rice sketchbooks
p. 151 Viviane Camerlo; Humbert Camerlo; Humbert Camerlo; Humbert Camerlo; Humbert Camerlo; Humbert Camerlo
p. 152 all pictures, Humbert Camerlo
p. 153 all pictures, Humbert Camerlo
p. 154 Humbert Camerlo; Humbert Camerlo; Alain Vivier; Jorge Damonte
p. 155 all pictures, Barbara-Ann Campbell

p. 156 all pictures, Andrew Sedgwick/OAP
p. 158 Nicolas Prouvé; Viviane Camerlo; Humbert Camerlo; Barbara-Ann Campbell
p. 159 Humbert Camerlo; Humbert Camerlo; John McMinn; Nicolas Prouvé
p. 160 Humbert Camerlo
p. 162 all pictures, Aerofilms Limited
p. 163 Dennis Gilbert; Ed Byrne; Ed Byrne; Ed Byrne; Patrick Lichfield
p. 164 Ed Byrne
p. 184 map designed by Kira Leroy and Antoine Baz

With thanks to:
Pauline Shirley, Jo Tomlinson, Trevor West – Ove Arup & Partners Slide Library
Isabella Carpiceci, Carla Gabato – Renzo Piano Building Workshop Archives
Karen Parker, architect

APPENDIX I

Peter Rice

Peter Rice, said Renzo Piano, the Italian architect with whom Rice collaborated on some of the most imaginative buildings of the past 25 years, designed structures 'like a pianist who can play with his eyes shut; he understands the basic nature of structures so well that he can afford to think in the darkness about what might be possible beyond the obvious.'

Rice was one of the most rigorous and imaginative structural engineers of his generation, much loved by the architects he worked with – most notably Sir Richard Rogers and Renzo Piano – and for whom he made possible some of the most unlikely and adventurous buildings of any age. Without Rice there would have been no Sydney Opera House (the first building he worked on, from the age of 23), no Centre Pompidou, no Lloyd's building and no Pavilion of the Future at Expo '92, Seville, the last finished design he worked on. Or at least, these designs would have turned out very differently without him.

In July this year, the British architectural profession paid homage to Rice by awarding him the Royal Gold Medal for Architecture, an accolade that rarely comes the way of engineers. One previous recipient was the brilliant Danish engineer Ove Arup with whom Rice made his career when he joined Ove Arup & Partners fresh from college in 1956. Although he was to set up his own practice, RFR, in Paris with Martin Francis and Ian Ritchie after the completion of the Centre Pompidou in 1977, Rice remained a hyperactive partner and director of Arup's, based in London.

Fond of poetry, philosophy, mathematics, racing, football (he was an avid QPR supporter), wild flowers, wine (which he collected) and whisky (Scotch before Irish), Rice was, perhaps, the James Joyce of engineering. His poetic invention, his ability to turn accepted ideas on their head and his rigorous mathematical and philosophical logic made him both one of the most sought-after engineers of our times and an inspiration to the legion of young engineers who followed in his wake. A tireless and often informal chatterbox, he was able to disarm the most pompous client and win over the shyest student.

Rice was born in rural Ulster, a place, he said, 'where architecture and engineering simply didn't exist'. A natural mathematician, he studied engineering first at Queen's University, Belfast and then at Imperial College, London. Taken on by Ove Arup, his first job was the seemingly impossible task of raising the gull-beak roofs of

An Engineer Imagines

Sydney Opera House designed by the Danish architect Jørn Utzon.

From then on his career was as meteoric as it was peripatetic, commissions taking him to Paris (Centre Pompidou, with Richard Rogers and Renzo Piano), Houston (Menil Art Collection Museum, with Renzo Piano), Bari (San Nicola World Cup Stadium, with Renzo Piano) and Japan, where he worked with Piano on the design of the new Kansai International Airport.

In Britain, his best-known works are the Lloyd's building (with Richard Rogers) and the new terminal at Stansted Airport (with Sir Norman Foster). His latest structure was the vertiginous stone arcade fronting the Pavilion of the Future at the Seville Expo. The free-standing filigree screen made an entirely new and daring use of stone, a material that, for Rice, had as lively a future as steel, concrete and textiles.

In the speech he made at the Royal Gold Medal ceremony in London in July, Rice said that structural engineers have been expected to play the role of Shakespeare's Iago, who undermined the love of Othello and Desdemona by reducing to reason their every unreasonable act or feeling. The engineer had been seen to reduce every unreasonable and soaring idea an architect might have. The true role of the engineer, said Rice, was not to reduce, but to explore materials and structures as had the great Victorian engineers and medieval cathedral-builders he so admired.

Yet although buildings Rice worked on – such as Lloyd's, the Pompidou Centre, the De Menil Museum, Stansted Airport and the new TGV stations for the SNCF – are imbued with the same art, logic and humanity that make Beauvais Cathedral or the Clifton suspension bridge so stirring, Rice and the architects he worked hand-in-glove with have yet to convince most builders, developers, princes and the sceptical British public that honestly expressed and supremely imaginative structures are superior to cheapskate buildings clad in corny Post-Modern or pseudo-historic fancy dress.

Rice lived his life at a tremendous pace; in recent years he spent a day a week in Paris and a day a fortnight with Renzo Piano in Genoa. A brain tumour began to slow him down a year ago and he was forced to stop jetting backwards and forwards to Japan where he was working on the construction of Kansai Airport. He worked for the last year of his life from his home in Shepherd's Bush, west London. A very private family man and a much-loved husband and father, he had previously kept the weight of his work well away from home. He carried on working, dreaming and exploring new ideas, as everyone expected him to do, until his death at the beginning of the week.

Jonathan Glancey

Peter Rice, structural engineer, born 16 June 1935, Ove Arup & Partners 1956–92, Director 1978–92, partner RFR (Paris) 1982–92, Hon Fellow RIBA 1988, Royal Gold Medal for Architecture 1992, author of *Le Verre Structurel* (with Hugh Dutton) 1990, married 1960 Sylvia Watson (one son, three daughters), died London 25 October 1992.

Reprinted with permission from *The Independent* Obituaries, 29 October 1992

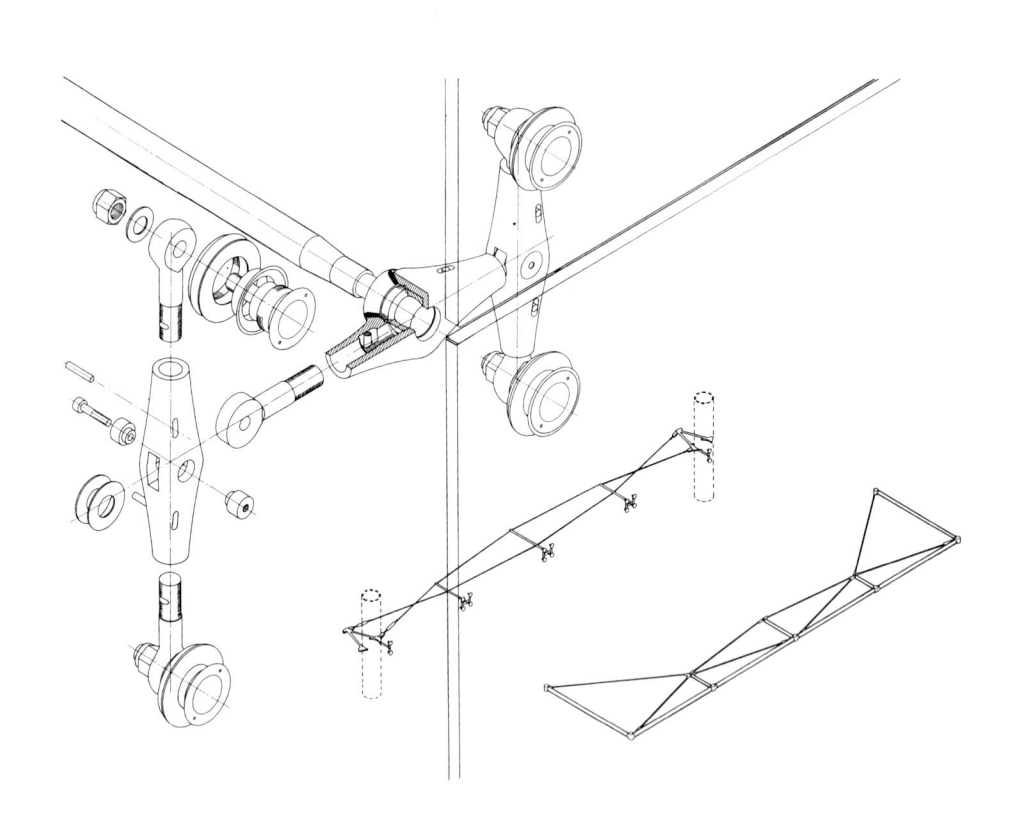

APPENDIX 2

RFR

RFR was started in 1982 as a response to a particular situation which arose at that time. After Beaubourg I had formed Piano & Rice with Renzo to try and combine architects and engineers in equal partnership. This did not work out, partly because the commissions were all architectural in nature, leaving no room for independent engineering commissions, and partly because of the difficulties of location. For all sorts of reasons it is impossible to operate independent personal practices in countries which have as diverse a view of tax and other legal liabilities as Italy and Britain, so the practice was discontinued.

At that time I was offered the opportunity to work with Adrien Fainsilber on a competition project he had just won for the design of the grand serres at La Villette in Paris. These were large glass conservatory structures. Because they were glass structures, I invited Martin Francis to join me and he invited along Ian Ritchie, an architect he had met while working at Foster Associates. Thus the team was formed.

Once established, the nature and character of what it could become became clear. Firstly, Henry Bardsley, a young engineer, and then Hugh Dutton, a young architect, were employed. The La Villette project was very successful and it was evident that the chemistry was working. After La Villette, Ian Ritchie withdrew while Martin Francis remained as a sleeping partner.[1] RFR continued as a laboratory interface between architecture and engineering.

At this point it started to become clear what the advantages were of having a small, inventive, innovative group to counteract and challenge the more formal way of thinking of larger frameworks like Arups. It was clear too that if one wanted to work on projects that would be built in France, they would have to be done by a firm based in France, dedicated to being French in every way. Being Irish I felt European, not British, so the transition to French working life was not a transition at all, just a natural progression of the time I had already spent in France, working on Beaubourg.

Once RFR was established its staff became predominantly French, with the clear belief that it was formed to work in Europe, and that was how we sought work and how we carried it out. Based in Paris, RFR is an engineering group but with the involvement of architects giving it design aspirations, typified by the La Villette façade. We realized that there were a number of clients, architects and others, seeking this kind of assistance. Engineering was always the driving force, with the emphasis, where possible, on invention and innovation, initially and particularly with glass.

[1] Martin Francis resumed active involvement in RFR after Peter Rice became ill in 1991.

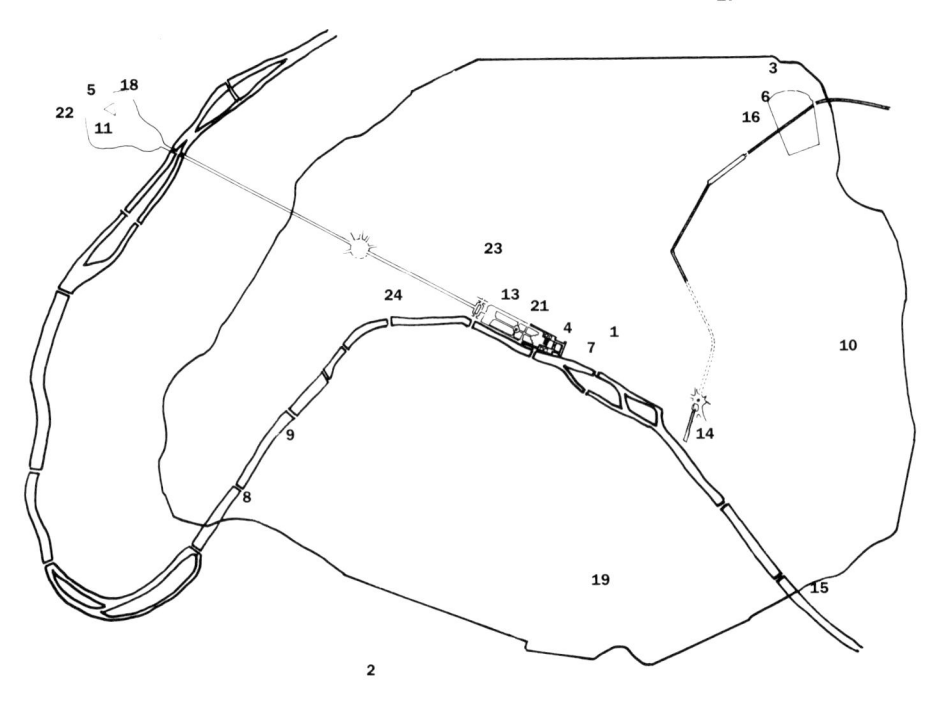

1 Centre Pompidou, 1971
ARCHITECT: Piano & Rogers
CONSULTING ENGINEER: OAP

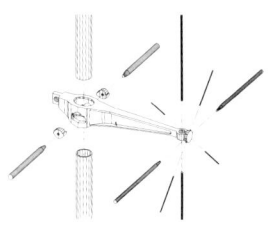

2 Schlumberger Headquarters roof, Montrouge, 1980
ARCHITECT: Renzo Piano Atelier de Paris
CONSULTING ENGINEERS: RFR in association with OAP

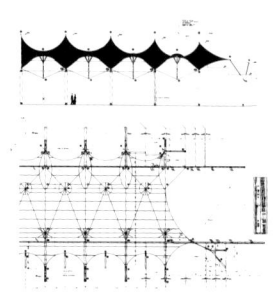

3 Serres and Toiture Accueil, Cité des Sciences et de l'Industrie, La Villette, 1981
ARCHITECT: Adrian Fainsilber
CONSULTING ENGINEER: RFR

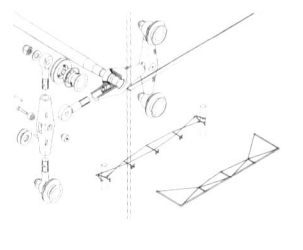

4 Louvre courtyard roof, 1985
ARCHITECT: I M Pei with Michael Macary
CONSULTING ENGINEER: OAP

5 Nuage Léger, Tête Défense, La Défense, 1986
ARCHITECT: J O Spreckelsen and Aéroports de Paris/Paul Andreu
CONSULTING ENGINEERS: RFR in association with OAP

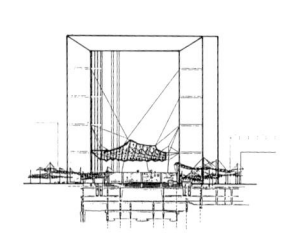

6 Galeries, Parc de la Villette, 1986
ARCHITECT: Bernard Tschumi
CONSULTING ENGINEER: RFR

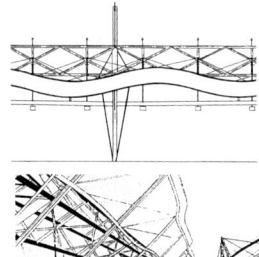

7 Passerelle Lintas, 22 Quai de la Magisserie, 1986
ARCHITECT: Marc Held
CONSULTING ENGINEER: RFR

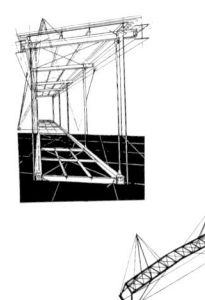

8 Parc Citroën Cévennes, greenhouses, 1987
ARCHITECT: Patrick Berger
CONSULTING ENGINEER: RFR

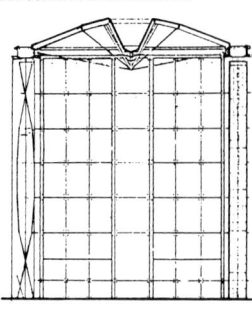

9 Passerelles Front de Seine, Beaugrenelle, 1987
DESIGN AND CONSULTING ENGINEER: RFR

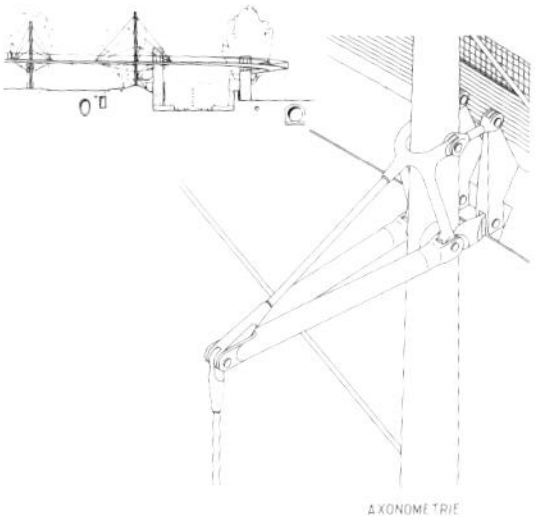

10 Atrium, Offices for Bull, Avenue Gambetta, 1987
ARCHITECT: Valode et Pistre
CONSULTING ENGINEERS: RFR in association with OAP

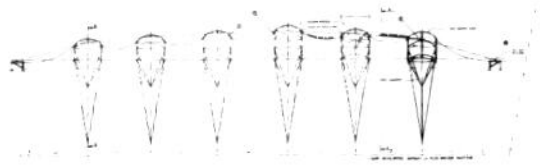

11 La Grande Nef, Tête Défense, La Défense, 1988
ARCHITECT: Jean-Pierre Buffi
CONSULTING ENGINEER: RFR

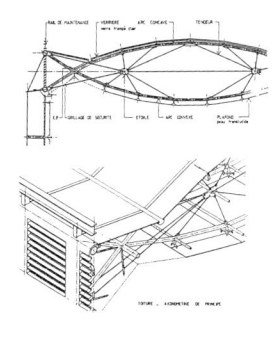

12 TGV/RER Charles de Gaulle, Roissy, 1988
ARCHITECT: Aéroports de Paris/Paul Andreu
CONSULTING ENGINEER: RFR

13 Les Tours de la Liberté, 1988
ARCHITECT: Hennin et Normier
CONSULTING ENGINEERS: RFR in association with OAP

14 Porte Monumentale, Quai de la Rapée, 1988
ARCHITECT: A Zublena
CONSULTING ENGINEER: RFR

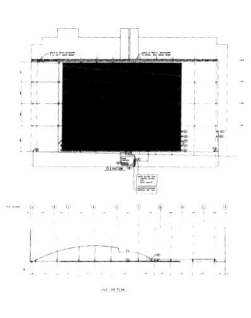

15 Bercy 2 roof system, 1988
ARCHITECT: Renzo Piano Atelier de Paris
CONSULTING ENGINEER: OAP

16 Passerelle Est/Ouest, Galeries Nord/Sud and bridge, Parc de la Villette, 1989
ARCHITECT: Bernard Tschumi
CONSULTING ENGINEER: RFR

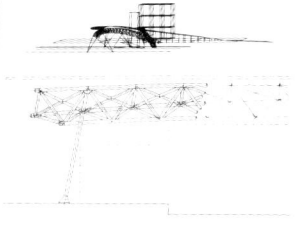

17 L'Oreal Factory, Aulnay, 1989
ARCHITECT: Valode et Pistre
CONSULTING ENGINEER: OAP

18 CNIT canopy and façades, La Défense, 1990
CONSULTING ENGINEER: RFR

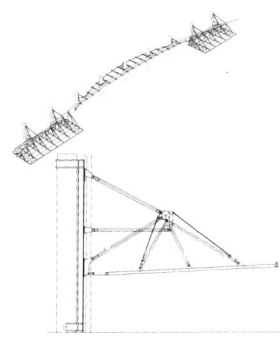

19 Place d'Italie, 1990
Atrium roof, Grand Ecran
ARCHITECT: Kenzo Tange
CONSULTING ENGINEER: RFR
Campanile
ARCHITECT: Kenzo Tange
SCULPTOR: Thierry Vide
CONSULTING ENGINEERS: OAP with RFR

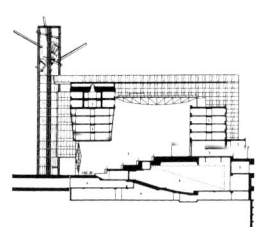

20 Terminal 3, Aéroport Charles de Gaulle, Roissy, 1991
ARCHITECT: Aéroports de Paris/Paul Andreu
CONSULTING ENGINEER: RFR

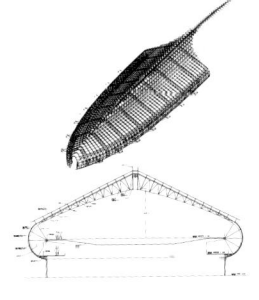

21 Pyramide Inversée, Grand Louvre, 1991
ARCHITECT: I M Pei
CONSULTING ENGINEER: RFR

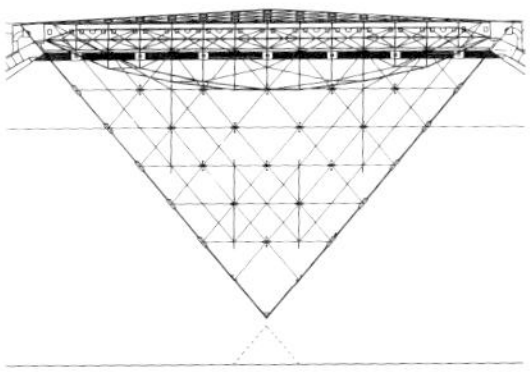

22 Japan Bridge, La Défense, 1991
ARCHITECT: Kisho Kurokawa
CONSULTING ENGINEERS: RFR in association with OAP

23 Ligne Météor, 1991
ARCHITECT: Bernard Kohn
CONSULTING ENGINEER: RFR

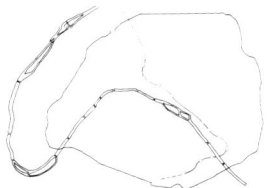

24 Atrium glazing, 30 Avenue Montaigne, 1991
ARCHITECT: O Vidal
CONSULTING ENGINEER: RFR

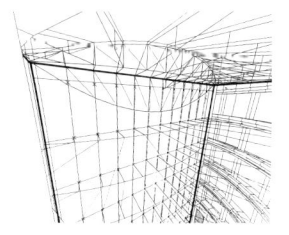

Index

Page numbers in *italics* indicate illustrations.

Abbott, Laurie 31, 32, 44
Aéroport Charles de Gaulle, Paris *185*
Aéroports de Paris 184, 185
Ahm, Povl 31
aircraft industry 82
Albert Dock, Liverpool (now Maritime Museum) *29*
aluminium
  cast 107, 112
  extruded 82
  and wood 65, *113*
analysis facilities 31, 112
  see also computer analysis
Andreu, Paul 100, 101, 184, 185
anti-clastic surface 96, 97, 99
aqueducts *29*, 120
Aquinas, St Thomas 52, 56
arches
  Centre d'Art Contemporain, Luxembourg 99
  MOMI Tent 99
  Pavilion of the Future 120, 122, 124, *124*, *125*
Archigram *26*
architect
  and the classical order 71
  co-operation with engineer 145
  creativity 72, 80
  and image-making 73
  as the main star 27
  and photography 127
  and specifications 45
  subjectivity 71
'architect engineer' 71
Architectural Association 25, *26*
Arctic City *96*
Arup, Ove 60, 67–70, *70*, 143
Ateliers de Gourgoubès 149
atrium,
  and fabric *105*
Auden, W H 75–6
Augustine, St 51–2
automobile industry 82, 135, 136
Avenue Montaigne, Paris *185*
axle,
  and gerberette 33

Bari Sports Stadium, Italy 97, *98*, 103
Bardsley, Henry 183

Barker, Tom 44, 88, *92*
beam nodes 38
beams
  cast-iron *29*
  double boom *79*, 79
  large-span 26
  primary *29*, 116
  reheating 41
  secondary *29*, 116
  and spherical bearings 33
  and standards 40
  suspended 32, 34
  transportation *40*
bearings,
  spherical *33*, 33
Beaubourg, Paris
  aerial view *24*
  budget 36, 37
  use of cast steel 29-31, *33*, 33, 46, 63, 77, 81, 90, 93, 115-16, *116*
  the client 34–5, 37, 40, 41, 46
  competition 25–9, 30, 35, 37, 38, 60, 81
  contract 37, 39, 44–5
  cross-section *28*
  importance of detail 28
  double boom beams *32*, *79*, 79
  façade 30, 32, 42, 44
  fire protection 26, 29
  flexibility 28–9
  floors 26, 30, 42
  framework 26
  front elevation *26*
  gerberette 32–4, 37–8, *38*, *39* 40, 41, *42*, *116*, 117, *184*
  movement zones 29, 32
  and photography 128
  piazza 29, 30, 44
  and politics 42
  renamed Centre Georges Pompidou 42
  scale of 28, 30
  services 29, 44
  span 28, 32
  standards problem 40
  structure *27*, 32, 33, 34, 37, 39
  teamwork 31
  tender 34–6, 39–40
  tests 37–8, 41
Belfast 57
Bennelong Point *62*, *63*
Bercy 2 (roof system) *185*
Berger, Patrick 184

Black and Tans 49
Blanchard, John 32
Bordaz, Robert 27, 34–5, 37, 42, *46*, 46
Bosphorus Bridge 74
bridges 79
  Bosphorus 74
  Clifton Suspension *77*
  Gerber *32*, 32
  Humber 74
  Menai Straits *77*
  Salginatobel *77*
  Severn *74*, 74
  stone arch 122
  suspension 74
  Verrazano Narrows 74
British Standards Institute 40
Brullmann, Cuno 44
Brunel, Isambard Kingdom 73, *77*, *77*, 79
Brunel, Marc 73
Buffi, Jean-Pierre 185
building industry 77, 133, 134
Bull Offices, Avenue Gambetta, Paris 104, *105*, 106, *185*
Bundoran 49
bureaux de contrôle 35, *113*
Burrell competition 26

cables
  and fabric 25, *98*
  in MOMI Tent *99*
Cahill (premier of New South Wales) 65
Camerlo, Humbert 149–50, *151*, 153, 154, 158, 159
cantilevers *32*, 32, 122
Carson, Willie *163*
cast iron *29*, 29, 90, 93, 94
cast steel
  and analysis facilities 31
  in Beaubourg 29-31, *33*, 33, 46, 63, 77, 81, 90, 93
  element of surprise and unpredictability 30
  and foundries 33
  and gerberette solution *33*, 63
  and load 33
  nodes *29*, *29*
  and personal contact 29, 46
casting
  and ductile cast iron 93
  and gerberette 32, 33, 34, *42*
  and standards 40

cathedrals,
    Gothic 29, 76, 78
Cauthen, Steve *163*
cement,
    ferro- *see* ferro-cement
Centre Beaubourg *see* Beaubourg, Paris
Centre d'Art Contemporain, Luxembourg 99
Centre Pompidou, Paris *see* Beaubourg, Paris
CEP 113
Challenger space shuttle 73
Chartres Cathedral 78
Chelsea football ground 25
Christian Brothers 50
Churchill, Sir Winston Spencer 50
City of London *162*
classical order 71
Clifton Suspension Bridge 77
CNIT, La Défense, Paris 83, *85, 185*
Collins, Michael 51
columns
    cast-iron *29*
    centrifugally cast 34
    and gerberette *32, 33, 42, 116, 117*
    nature of 32
    Pavilion of the Future 122, *123, 125,* 129
communication 27–8, 40, 42, 44–5, 53, 68, 131–2
computers 81, 89, *92,* 94, 95, *96,* 102–4, 122, *124,* 124, 143, *158*
concrete
    and bridges of Maillart 77
    Lloyd's of London *72,* 72, 115, 116, 117
    precast 94, 116
    properties of 72
    and Sydney Opera House shells *63,* 64
    tubes used in Conway Crossing 74, *74*
construction,
    engineers' work with 76
Conti, Mr (of Fiat) *136*
contractors 44, 45, 124–5
Conway Crossing 74, *74*
cooling, natural 83
Cornell University 65–6
County Louth 49
Crystal Palace, London 134

Cube Arch, La Défense, Paris *see* Grande Arche, La
culture 30, 45

Daussy (chief engineer of Socotec) 35
Davies, Michael 44
Day, Alistair 124
Dekany, Andrew 31
Derby, the (Epsom) 162, *163*
designer
    and image-making 73
    and the material 77–8
    subjectivity 71
detail, importance of 28
Dowd, Mike 135
dress-designing *104,* 104
Dublin 50, 52
Dundalk 47, 49, 50, 51, 55
Dutton, Hugh 183
Dyer's Hand, The (Auden) 75

Eddery, Pat 161, *163*
Eiffel, Gustave Alexandre 73
Eiffel Tower, Paris 26, 30
Elphick, Mike 59
energy, conservation of 83
engineer
    sense of adventure 31, 77
    co-operation with architect 145
    and economy 78
    as entrepreneur 73
    excitement of work 74–5
    exploration 77
    image of 72–3
    'Iago mentality' 75–6
    innovation 74, 75, 77, 78, 80, 143
    inventiveness 72, 74, 75, 143
    nature of profession 30–1, 67
    need for identity 74
    in nineteenth-century 73, 81
    objectivity 71–2
    role of 71–80, 81–2, 131
    and safety 75
    working in concepts 79–80
environment 68, 78
EPAD 102
Esch, Luxembourg *106*
établissement public 37
Expo '92 Committee 119

fabric
    as a cheap solution 106

and computer technology 102–4
cutting patterns 97, 102, 103
and glass 104, *106*
as a highly specialized subject 96
lack of presence 99, 100
modelwork 95–6
overlapping 104, 106
and polycarbonate 104, *105,* 106
prestressed 95, 96
seaming 97, *97,* 98
as a translucent material 97, 101, 106
Fainsilber, Adrien 108, 145, 183, 184
Ferrari Coupe *71*
ferro-cement 87–8, *89,* 90, *91, 92,* 93, 94
fettling 38
Fianna Fail 51
Fiat project 79, 135–42, *143*
Fiat Ritmo *136*
finite element analysis 95
fire protection 26, 29
Fleetguard factory, Quimper *128,* 128
floors 26, 30, 42, 79, 116, *117*
football 161
forces
    and fabrics 95
    and the gerberette 32–3
    primeval 74
foundries
    and cast steel 33, 37
    craft-based methods 33
    and the gerberette 32, *38, 39*
fracture mechanics 33, 40, 41, 44
Francini, Gianni 44
Francis of Assisi, St 52
Francis, Martin 183
French government,
    and Beaubourg competition 25, 27, 35
Frost, Robert 163
Full-Moon Theatre, Provence 149–60
Future Systems 99

Gaelic football 48-9, 161
Galerie, La Villette *147*
Galois, Evariste 52
Gare de Lyon, Paris 26
General Electric 141
Gerber, Heinrich 32
gerberette solution 32–4, 37–8, *38, 39,* 40, *41, 42,* 63, *116,* 117
girders,
    and Menil Museum *92, 93*
Giscard d'Estaing, Valéry 42

Index

Giugiaro 71
glass
   CNIT façade *85, 185*
   La Villette 'greenhouses' 72, 78, *184*
   and Menil Museum 88
   and Nuage Cube *101*, 101, *184*
   properties of 72, 108, 119, 145
   strength of 108
Gothic architecture 63
   cathedrals 29, 76, 78
Grand Palais, Paris 26
Grande Arche, La Défense, Paris, La *100*, 100, 101, 102, *185*
Grande Bigo, Expo '92, Genoa, Il *106*
Grandes Serres, La Villette, Paris 72, 107–9, *109–11*, 112, *113, 114*, 145–6, *146, 147*
granite 120, 126
Groningen 23
Grut, Lennart *31*, 31, 32
GTM 36, 44
Gyles Quay 47, 49, 55, 56

Happold, Ted 25, 27, *31*, 31
Hartley, Jesse 29
Hassfurt (Gerber bridge over the Main) *32*
Held, Marc 184
Hemingway, Ernest 67
Hennin et Normier 185
Heyman, Professor Jacques 122, *124*
Holt, Eric 44
horse-racing 161–2, *163*
Hostaflon plastic 99
Houses of Parliament 26
Houston 87, 88, 89, 131
Humber Bridge 74

'Iago mentality' 75-6
IBM Travelling Pavilion *107*, 107, *108, 109*, 112, 113
IDEA (Institute of Development in Automotive Engineering) 139
Imperial College, London 59
industry
   aircraft 82
   automobile 82, 135, 136
   building 77, 133, 134
   communication with 44
   French 34, 35, 36, 37, 81
   and gerberette solution 34
   prejudices of 112

responsible only to itself 142
role of engineer 77
role of 78
rolling investment programme 142
steel construction 36
working with 133-4
Inniskeen 47, 48, 49
innovation, of engineer 74, 75, 77, 78, 80
Institut de Soudure 34, 40
insulation 83
'Interchange' 26
inventiveness 72, 74, 75
IRA 50, 57
IRCAM 44
Ireland
   charm in 65
   and culture 30
iron
   cast *see* cast iron
   ductile 87, 90, *92*, 93, 94
   Telford's use of 77
Ishida, Shunji *89*, *92*, 135
Issigonis, Alex 142

Japan Bridge, Paris *185*
Jenkins, Ronald 59
Johnson, Philip 37
jointing
   and cars 137
   cast 29, 42
   and concrete 116
   epoxy 120
   expressiveness of 26
   and moveable floors 26
   and polycarbonate 107, 109
   and stone arch 122, *123, 124*, 126
   and tiles 28, 62
'Joker in the Pack, The' (Auden) 75–6

Kahn, Louis 116
Kandinsky, Wassily 104, *105*
Kansai Airport terminal project 36
Kaplicky, Jan 99
Kavanagh, Patrick 47, 48
Kelly, W Paul *92*
Kilkenny 49
Kingsley, Ben *76*
Kohn, Bernard 185
Krupp 36, 37, 39, 41, 44
Kurokawa, Kisho 185
Kussmaul, Professor 41
Kuwait Sports City *96*

La Villette, Paris 72, 78, 113, 119, *184, 185*
   see also Grandes Serres
labour costs 88
light
   and double booms of Beaubourg 79
   engineers' work with 76
   and Full-moon Theatre 149–59
   and Menil Museum 87, 88, *89*, 89, 94
lighting diffuser *106*
Ligne Météor (Paris) *185*
Linlithgow, Scotland 29
Lloyd's of London 72, *72*, 79, *115*, 115, 116, *117*, 117
load
   and computer modelling system 122
   eccentric 115
   environmental 147
   and fabric 95, 96, 101
   and glass 108, 109, *112*, 112
   and polycarbonate 107, 109
   and steel 35, 115-16
   and stone 119, 122
   wind 122, 147
load-testing 35–6, 37–8, *113*
London School of Economics 49
Lord's Cricket Ground, London 97, *98*, 103
Los Angeles *134*
Louvre, Paris 30, *185*

Macary, Michael 184
machining
   and gerberette solution 34
   and re-heating 41
Mackay, David 119
Mackenzie, Ian 59
Maillart, Robert *77*, 77
Maison Coloniale 82
Maison du Peuple, Clichy 83, *84*
Maison pour le prospecteur solitaire du Sahara (Prospector's Hut) 83, *84*
Maison Tropical, Niamey 82, 82–3, *83*
Mantagazza, Francesco 139
Martorell Bohigas Mackay (MBM) 119
Masonry Arch, The (Heyman) *124*
masts,
   and fabric *97*, *98*
materials,
   engineers' work with 76–8
mathematics 48, 52, 53, 57
MBM *see* Martorell Bohigas Mackay

media,
    and the image 72–3
membrane 96, 97, 98
Menai Straits 77
Menil Collection Museum, Houston, Texas
    aerial view 86
    and ductile iron 90, 92, 93, 94
    and ferro-cement 87–8, 89, 90, 91, 92, 93, 94
    and glass 88
    and light 87, 88, 89, 89, 94
    and louvre system 87, 88–93, 88–90
    plastering 87, 88
Menil Foundation 88
Menil, Mrs Dominique De 87, 88, 89, 90
mesh
    and ferrocement leaves 90
'finite element' 96
Mini, the 142
modelwork 95–6
Modern Movement 68
MOMI (Museum of the Moving Image) demountable tent, South Bank, London 99
Momix 154, 154
moon reflectors 151, 151, 152, 153, 153, 154, 156, 158, 158, 159, 159
Moon Theatre see Full-Moon Theatre
movement zones,
    in Beaubourg 29, 32
Museum of Modern Art, Paris 37

Nanyang project, Singapore 104, 105
National Gallery, London 30
Nervi, Pier Luigi 87, 125
nets 95, 124
New York 134
Newbridge 57
Nippon Steel 36
Noble, Neil 108
Notre Dame Cathedral, Paris 78
Nuage Cube, La Grande Arche, La Défense, Paris 100-102, 104
Nuage Parvis, La Défense, Paris 102, 102–3
Nuit des Etoiles Filantes, Les (television presentation) 154

Oberhausen station square competition 106
O'Brien, Turlogh 32

oil platforms 33, 40
Okabe, Noriaki 135
Old Man and the Sea, The (Hemingway) 67
Oreal, L', factory 185
Orly DC10 aircrash (1974) 41, 73
Osaka World Fair (1970) 29, 29
Othello (Shakespeare) 76, 76
Otto, Frei 25, 66, 95, 96
out-of-plane loading 122
Ove Arup & Partners 27, 156, 183
    critical role of back-up team 31–2, 124
    Peter Rice joins 59, 67
    Structures 3 group 25, 26, 66

Palacio da Ajuda, Lisbon 119, 119
Parc Citroën Cévennes, Paris 184
Paris
    map of Peter Rice's buildings and projects 184
    metro stations (Art Nouveau entrances) 26
    steel structures in 26
Paris see also under individual buildings
Passerelle Lintas, Paris 184
Patscenter project, Princeton, New Jersey 129, 130
Pavilion of the Future, Expo '92, Seville 119–26, 129, 131, 131
Pei, I M 99, 184, 185
Pendleton, Moses 154
PHB see Polig Heckel & Bleihart
photography 127–32
Piano & Rice 31, 135, 183
Piano & Rogers 25, 26, 28, 30, 60, 185
Piano, Renzo 26
    and Beaubourg 25, 31, 46
    and Fiat project 135, 136, 139, 140, 141
    and IBM Pavilion project 107
    and Menil Museum project 87, 88, 88, 89, 90, 92, 93
    partnership with Rice 135, 183
Pininfarina 71, 71
plastering (ferro-cement) 87, 88
plastic
    Hostaflon 99
    mirrored 151, 155
Plattner, Bernard 135
Plunkett, Oliver 52
plywood
    and Moon Theatre 151, 152, 153, 155
    reinforced with aluminium 65

poetry 54
Polig Heckel & Bleihart (PHB) 37, 41, 44
politics
    and Beaubourg 42
    and Sydney Opera House 60
polycarbonate 107–12
polytetrafluoroethylene (PTFE) see Teflon-coated glass fibre
Pompidou Centre, Paris see Beaubourg, Paris
Pompidou, Georges 37, 42
post-modern style 131
prestressed fabric 95, 96
Prospector's Hut (Maison pour le prospecteur solitaire du Sahara) 83, 84
Prouvé, Jean 37, 81–5, 125
Provisional IRA 50
PTFE (polytetrafluoroethylene) see Teflon-coated glass fibre
pultrusions 99, 99
PVC 97, 100

Quai de la Rapée, Paris 185
Queen's University, Belfast 57
Quinn, Dada 47, 48

Ravensdale 54
Raymond, Bruce 163
RFR 85, 124, 183–185
RIBA see Royal Institute of British Architects
Rice, Nemone 27, 162
Rice, Peter
    childhood 47–57
    and the Church 56–7
    education 27, 50–51, 53, 57, 59, 65–6
    and formation of Piano & Rice 135
    idea of using cast steel for Beaubourg 29–31
    joins Ove Arup & Partners 59, 67
    learns about architecture 59–60
    and mathematics 48, 52, 53, 57
    and poetry 54
Ritchie, Ian 183
rods,
    in tension 42
Roissy (TGV–RER station) 185
Rogers, Sir Richard 25, 26, 31, 46, 79, 116
Royal Institute of British Architects (RIBA) 26

Index

Royal Shakespeare Company 76
Royal Ulster Constabulary 47

Saarinen, Eero 60
SAEM 102
Salginatobel Bridge, Switzerland 77
scale,
　importance of detail in determining 28
Schlumberger Headquarters,
　Montrouge 97, *97*, 99-100, *184*
Sedgwick, Andy 156
Severn Bridge 74, *74*
Slammannan Railway *29*
soap film surfaces 95, *96*
Socotec 35, 40, 113
specifications 45
Spence, Robin 26
Spreckelsen, J O 100, *101*, 102 184
spring, prestressed 109, 112
standards
　British 40
　French 39–40
　German DIN 40
Stanton, Alan 44
Stanton, Johnny 31, *32*
steel
　Brunel's use of 77
　cast *see* cast steel
　pressed *138*
　sheet 82, 83, *84*
　uniformity of standard sections 30
steel construction industry 36
steel structures, in Paris 26
Stephenson, Robert 73
Stokes, Father 50
stone
　cutting 120, 125–6
　physical presence of 76
　prefabricated 120
　properties of 119
Structures 3 group *see under* Ove
　Arup & Partners
structures
　lightweight 95
　in nature *95*, 95
　prefabricated 82–3
structures, design of large 74
Stuttgart University
　Institute for Lightweight Structures *96*
　Institute of Materials 41

Suchet, David 76
Sunday Times Insight team 41
Sydney Harbour Bridge 64, *66*
Sydney Opera House 27, 30
　the client
　competition 28, 60
　importance of detail 28
　floor plan *60*
　free-spirited concept *58*
　integrity 63, 65
　podium 59, *60*, 60
　and politics 60, 63, 65
　precast elements 59
　and quality 62
　resignation of Utzon 63, 65
　rib sections 59, *62*, 62, *63*, *64*
　roof structure 59
　scale 28, 63
　shells 59, *60*, 60, *61*, 62, 62, *63*, 63, *64*, *65*
　as symbol for Sydney 28, 65
　tiling 59, 62-3, *63*, *64*, *65*, *66*

tactility of buildings 76-7, 78
Tange, Kenzo 29, 185
Tate Pavilions, London *104*
teamwork 31, 73
Teflon-coated glass fibre
　(polytetrafluoroethylene; PTFE)
　96-7, *97*, 99, 100
Telford, Thomas 73, *77*, 77, 79
Tenara 99
tendering 34-6, 39–40
tents 95, *97*, 97, 124
Teresa, St 52
textile mill *29*
Thaury, Jean 44
Théâtre de la Pleine Lune 150
Thomson factory research facility,
　Conflans Ste Honorine 97
tiling 28, 59, 62-3, *63*, *64*, *65*
timber
　and IBM Pavilion 107, 109, *110*, *111*
　and Full-Moon Theatre 151, *155*
　properties of 112
Tokyo *134*
Tours de la Liberté, Les, Paris *185*
traces de la main 63, 76, 77
transparency,
　and La Villette 'greenhouses' 72
trusses
　at Menil Museum 88, *92*

cable *101*, 101, 109, *112*, *113*, 145
　and Grandes Serres 109, *112*, 145
　and Nuage Cube *101*, 101
　Pavilion of the Future *123*
Tschumi, Bernard 108, 145, 146, 184, 185
Tsuboi, Professor 29
tubes
　in compression 42
　concrete 74
Turkish Airlines 41

Utzon, Jørn 28, *58*, 59, 60, *62*, 62, 63, 65

Valode et Pistre 185
ventilation 83
Verbizh, Reiner 135
Veritas 113
Verrazano Narrows Bridge 74
Verre Structurel, Le (Rice) 108
Vidal, O 185
Vide, Thierry 185

Weinstein, Richard 76
wind-tunnel effect 101
Winkler, Paul 88, *89*

Yale University Art Gallery 116
York Minster (Rose Window) 112, *113*
Young, John 44

Zublena, A 185
Zunz, Sir Jack 59

Each time you go out you discover something which gives you another direction to turn to but of course there's another big problem. People say to me again and again, 'Oh, you've had such wonderful projects, oh my goodness you have had so many interesting projects, how can you be so lucky? Why don't I get half or some of those projects, why doesn't anyone else get them, why do they always come to you?' But in a way this is the point: although you have to have commissions, you have to make what you can out of the commissions you get. And, you know, gradually people come to you to buy surprise and the thing that's nicest about it is that when people come to buy surprise I have no idea of what I'm going to give them either. It's not like I'm going out of my way to surprise them, I'm actually quite often surprised myself by what the outcome is because I'm a bit like a hound following a fox; I'm following something really close to the ground and I can't actually see where it's going. I've got my nose to the ground to make sure I'm following it properly.

Peter Rice, 4 February 1992

An Engineer Imagines

Street Lights
The lightest-possible mast with a tuned mass damper at its top and suspended rings allowing many permutations of lamps at six, nine and 12 metre heights. Highly tensioned wire spokes maintain the ring's stability.

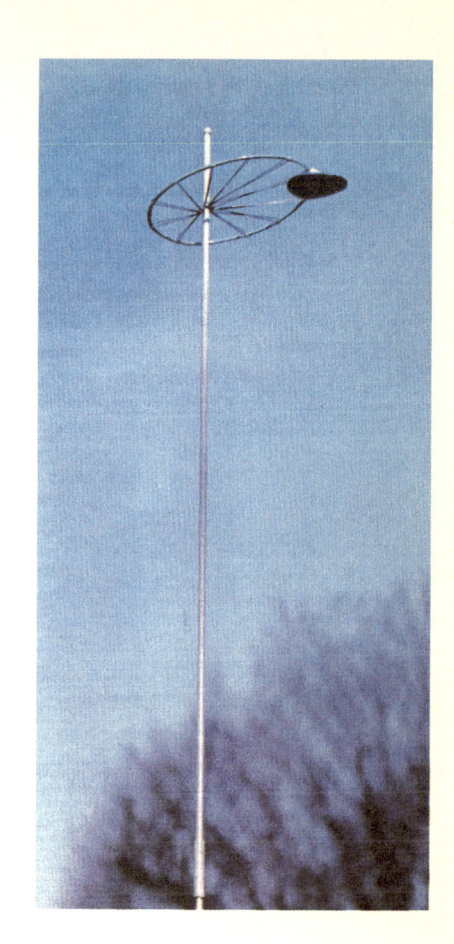

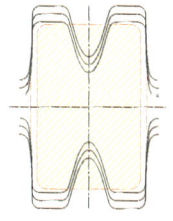

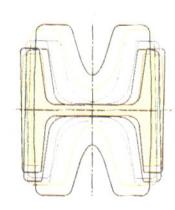

Heavy profiles at Esch
Lighting masts for the home of Arbed, one of the world's largest steel profiles manufacturer. The standard production process is interrupted, resulting in a curious 'dogbone' profile. Perforations at the top of the mast prevent aerodynamic excitation by interrupting air-flow paths.

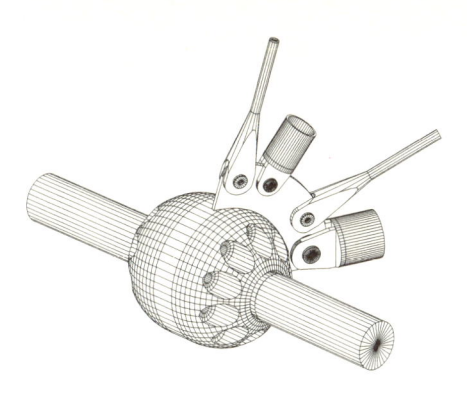

Japan Bridge
The concept of a tied arch makes the bridge very stable for uniform vertical loads but the relatively high arched and glazed walkway makes it sensitive to overturning when subjected to lateral wind loads. In order to resist this, the arch is twinned to form a triangle in section for overall loads, and the walkway, with its legs and tie cross-bracing, forms a torsion box for local and partial lateral loadings.

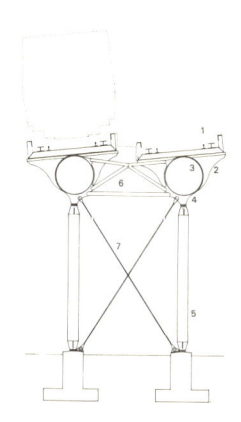

Riding the tube at Lyon
Elevated metro systems have a strong presence in our cities. In this case, the object is to express and articulate clearly an efficient structural system. The proposal supports each set of rails on bands of concrete which, in turn, rest on exterior 'wings' attached to a large-diameter steel tube with internal diaphragms.

An Engineer Imagines

Kansai
Renzo Piano imagined 'a giant plane alighting' on the densely planted artificial island. The roof form, shaped by the aerodynamics of the large-scale air jets ventilating the whole space, is rationalized for repetition and constructional logic. The roof hovers above the more static building levels, barely connected to them by slender grazing supports and leaning props. The lightweight tubular steel structure – three dimensional truss-arches and the 1.6 kilometre long tie-stiffened shell – is detailed to evoke early flight technology and to provide transitions in scale to people who use it.

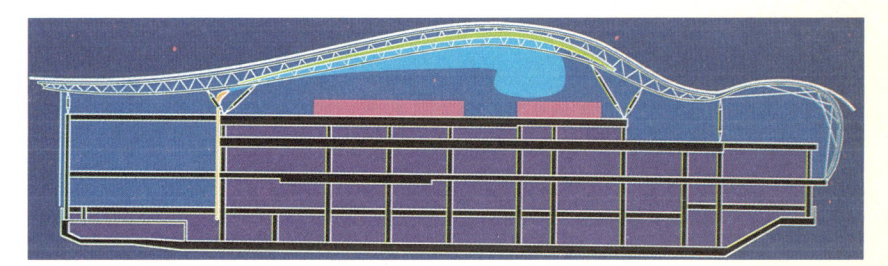

A floating roof at Lille
The roof is a vast luminous wave hovering over tie-stiffened steel arches. Continuity between inside and outside is maintained by the façades, treated as partial glass screens independent of the roof.

Steel croissants at Roissy
The roof structure of the new station serving Charles de Gaulle Airport is a transposition of an image of 'levitation'. Each structural system is expressed as an independent layer and is completely detached from the façades. Fritted glass panels appear to float on articulated struts which are supported on the croissant beams – a strong curved double bottom boom and a thin tensile top chord. These in turn are supported on fan-shaped pylons.

Seeing the mountain before you ski
A tubular steel arch structure supporting a glass roof at Chur, the last mainline rail station for the Swiss ski resorts. Each arch has two members which diverge and reconverge. The arches are stabilized from deformation by a fan of ties and are suspended from small stub columns at each end. The glazing is expressed as a distinct layer with a complementary geometry.

An Engineer Imagines

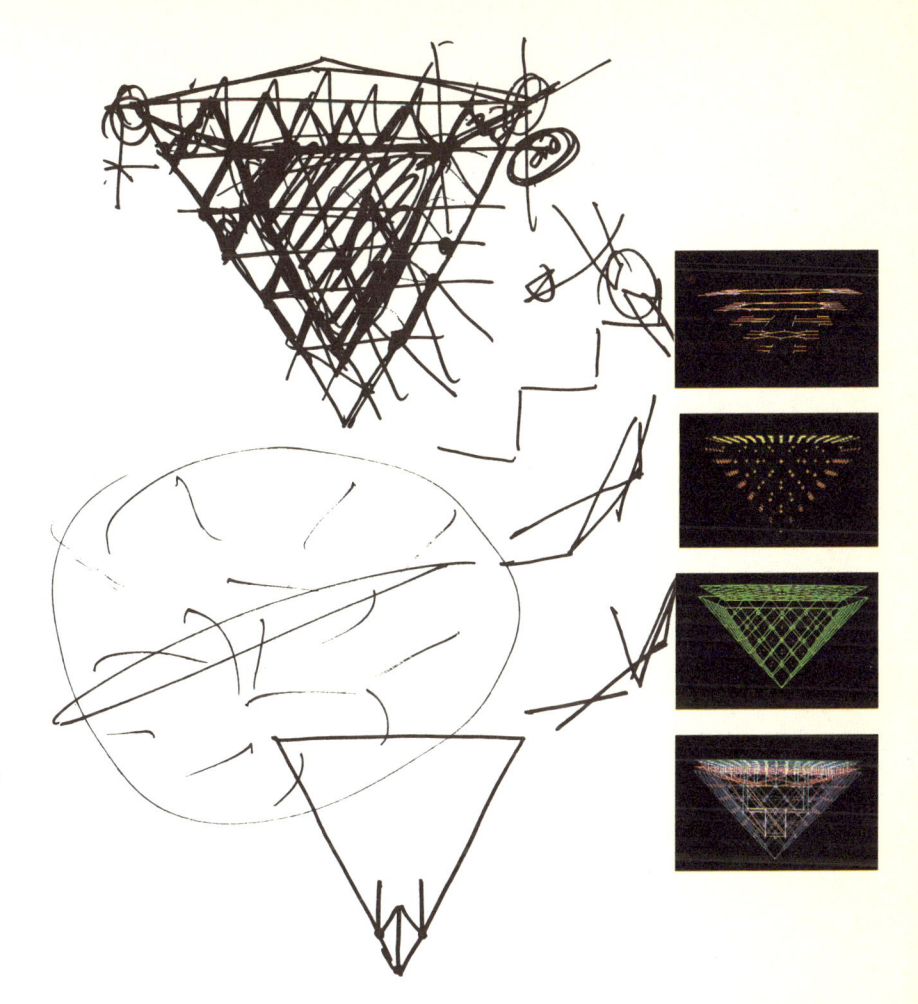

Inverted pyramid
The complexity of a structural system that is momentarily simple in the symmetry of the pyramidal glazed surfaces. The roof is slightly facetted; the triangular glass chandelier is hung from a primary structural frame. The suspended ties and struts, prestressed by the weight of the glass, balance the components of weight between opposite sides, and give geometrical contrast.

New entrance for the CNIT
Every piece of Prouvé's original façade has a specific function and its shape is determined by the way it works. The new design consists of a curved torsion tube cantilevered off the existing concrete structure from a system of steel collars slicing through the slender Prouvé mullions. The new design attempts to resolve a different problem in a different epoch whilst sharing a design philosophy.
An entrance canopy and glazed café roofs are fixed to the tube by a series of struts whose variable geometry is generated by projecting the curve of the tube down to a simple rectangular box.

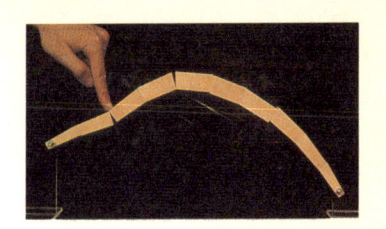

Padre Pio pilgrimage church
A series of natural stone arches provides a distinct primary structure on which a secondary roof structure is supported. The stone's strength is used by finding a geometry that allows load transmission primarily by compressive thrust.

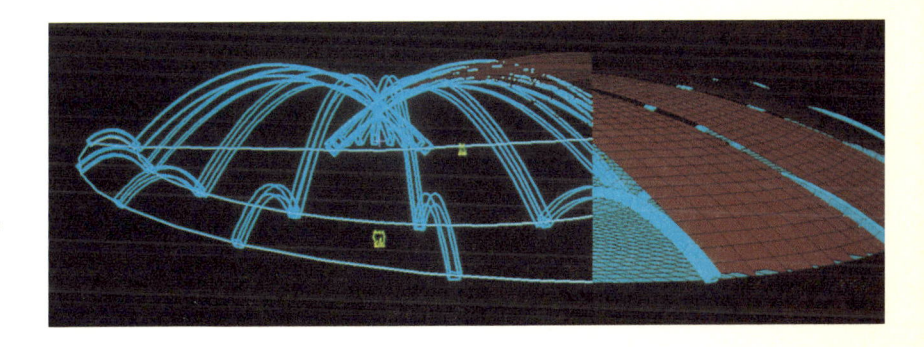

Byzantine frescoes
Thirty-eight fragments had to be pieced together. Computer graphics were used to recreate the geometric forms of the Cypriot church dome and apse from which they came. A supporting structure, a GRP/foam sandwich fabricated by boat builders, allowed the frescoes to be manœuvred and eventually transported without damage.

PUBLISHER'S NOTE

An Engineer Imagines was prepared for publication after Peter Rice's death. Great care has been taken to follow the author's wishes. The images and captions throughout the book have been selected and completed by many of the engineers and architects who worked with Peter Rice.

Patrick Kavanagh's poem on page 47 is reprinted by kind permission of the Estate of Patrick Kavanagh, c/o Peter Fallon, Loughcrew, Old Castle, County Meath, Ireland.

Published by Artemis London Limited
37 Alfred Place, London WC1E 7DP

© 1994 Sylvia Rice

All rights reserved. No part of this publication may be reproduced in any form without written permission from the Publisher.

British Library Cataloguing in Publication
A CIP record for this book is available from the British Library

ISBN 1 874056 21 8 (London)
ISBN 3 7608 8408 3 (Zürich)

Project editor Barbara-Ann Campbell
Designed by Jonathan Moberly with Kevin Whelan
Imageset by The Alphabet Set, London
Printed and bound in Hong Kong